Enigma
Of the Axial Age

Enigma
Of the Axial Age

History, Evolution
And the Macro Effect

Secularism, Modernity,
And the Evolution of Religion

An Abridged Edition of
World History and The Eonic Effect

John C. Landon

South Fork Books

© Copyright 2016
John C. Landon
Published by South Fork Books
Montauk, NY
Library of Congress Control Number: 2015919554

ISBN 978-0-9847029-4-7
Printed in the United States of America

TABLE OF CONTENTS

Preface   7
Introduction   49

2. The Axial Age: A Riddle Resolved   89

3. Conclusion: The Great Transition   241

4. Appendix: An Idea from Samkhya   260

5. Bibliography   277
6. Illustrations   287
7. Index   292

PREFACE

This is first edition of *Enigma of the Axial Age* (ENAX), an abridged version of *World History and the Eonic Effect*(WHEE), and an attempt to bring some clarity to the hopeless confusion over that period, via an introduction to the so-called 'macro' or 'eonic effect'. Many readers were unaware that the solution to the riddle of the Axial Age was in a book without that term in the title. The book closes a trilogy with companion volumes *Descent of Man Revisited* (DMR), and *Last and First Men* (LFM), and will accompany an abridged and revised fourth edition of *World History and The Eonic Effect* (WHEE). The book shows a 'triple play', an elegant connection of the enigma of the Axial Age, to the idea of evolution, and what we call Kant's Challenge.

The Axial Age is the master clue to the hidden structure of world history, the 'eonic' or 'macro' effect, and this in turn connects to the idea of 'evolution'. Nothing in world history makes sense without seeing the macro effect and that starts with seeing the Axial Age, a problematic term. We are moving to transcend the term altogether.

> It was the biologist Dobzhansky who noted that 'nothing makes sense' except in the light of evolution. But the corollary we suspect

is that nothing makes sense in the light of natural selection and the perspective of random evolution. It is suspiciously the kind of theory those who have never truly observed evolution might adopt.[1]

Fig. 1 The Triumph of Civilization, a pantheon of gods, 1793

How does 'evolution' enter the discussion? Because world history shows evidence of a developmental sequence. Our usage of the term is thus novel, but actually the right one, beside darwinian confusion over 'random evolution'. We explore the connection between history and evolution in a simple and elegant deduction of the finite transition model, and this shows what is happening with 'axial ages'. The term 'evolution' invokes a formalism of 'macro' and 'micro' as a two-layer developmental process, in which the Axial period is a stage. A critique of darwinian confusions is the starting point, along with the realization the Axial Age shows us the key to the evolution of religion.

The term 'eonic effect' has been extended, if not replaced, with the equivalent 'macro effect'. 'Macro' and 'micro' are category distinctions (they arise naturally in economics) and are definable in multiple contexts, here in evolutionary and historical terms both of which are related. We see that 'punctuated equilibrium' (as if two words fresh from the dictionary) is just such a category jargon, and is almost like a 'principle of sufficient reason'. But there again the jargon throws light on the Axial Age.

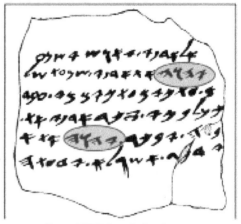

Fig. 2 IHVH on Lakis Letters

The *Introduction* starts with Kant's Challenge, and the critique of darwinism, while *Chapter 2* is the core with everything needed to understand

---

1 "Nothing in Biology Makes Sense Except in the Light of Evolution", American Biology Teacher, Vol. 35, pp. 125-9.

Fig. 3 Karl Jaspers

We start deliberately with an indication of the apparent theological cast of the opening to Jaspers' text. But students need to be wary here: such a thinker is steeped in German Classical Philosophy, with a trace of Kierkegaard, and is navigating the world of successors to this such as Nietzsche and Heidegger. These are his peers, and he is really a thinker able to stage a challenge to the Weberian mindset developing around him, save only that this pits the issue of 'faith' against an expanding view created by globalization. This allows someone closer to the mood of science to bring in the issue of teleology.

THE AXIAL PERIOD: "...from OGH...

In the Western World the philosophy of history was founded in the Christian faith. In a grandiose sequence of works ranging from St. Augustine to Hegel this faith visualised the movement of God through history. God's acts of revelation represent the decisive dividing lines. Thus Hegel could still say: All history goes toward and comes from Christ. The appearance of the Son of God is the axis of world history. Our chronology bears daily witness to this Christian structure of history.

But the Christian faith is only one faith, not the faith of mankind. This view of universal history therefore suffers from the defect that it can only be valid for believing Christians. But even in the West, Christians have not tied their empirical conceptions of history to their faith. An article of faith is not an article of empirical insight into the real course of history. For Christians sacred history was separated from profane history, as being different in its meaning. Even the believing Christian was able to examine the Christian tradition itself in the same way as other empirical objects of research.

> The opening to
>
> *The Origin and Goal of History*
>
> We are thrust 'sink or swim' by Jaspers into a subject that requires archeological science, and...a teleological question. Can we wrest a solution to this beautifully staged theoretical curve ball?

the Axial Age squashed together in one place to make the resemblance of the modern and Axial transitions obvious. The result shows why the problem of historical dynamics remains elusive: people don't read enough books, the data is too massive, and, as important, the data wasn't there until the nineteenth century onward. They are also conditioned against historical discontinuity, an idea subject to many confusions. But the Axial Age shows that such a phenomenon exists, and a simple model can relieve thinking from the confusions of darwinian flatland logic.

Fig. 4 Buddha ('Vajra Mudra')

It is very difficult to master such a huge data set, a subset of a still larger one. Our model is actually helpful here, and can assist visualization, if such a term is possible for large swaths of history. Be wary of the simplifiers. And the standard academic literature on the Axial Age is almost a joke on standard academic scholars. Our model can be assembled and dismantled at will. It helps to some clarification, but you are free to take a skeptical approach. I think the model will prove so useful that skepticism will alternate with some cautious use of it. We should note that the biological community has thought itself free to deceive the public about darwinism, so we should applaud our own work as outsiders.

Notes

Fig 5 Lao Tzu

The Axial Age is a mystery that summons all the others, that of Kant's challenge, that of a science of history, that of the evolution of religion, and that of the nature of monotheism and the nature of the spiritual. But it also shows the birth of the secular. The issues of historical analysis have been quarantined from issues of consciousness, and issues of divinity. The former are now a New Age obsession, while the latter, in the era of Kant, is the mirror image of still another issue, that of no divinity. The issues of the current 'new atheism' movement, and the confusions of secularism require our extension as the

Preface

> An axis of world history, if such a thing exists, would have to be discovered empirically, as a fact capable of being accepted as such by all men, Christians included. This axis would be situated at the point in history which gave birth to everything which, since then, man has been able to be, the point most overwhelmingly fruitful in fashioning humanity; its character would have to be, if not empirically cogent and evident, yet so convincing to empirical insight as to give rise to a common frame of historical self-comprehension for all peoples–for the West, for Asia, and for all men on earth, without regard to particular articles of faith. It would seem that this axis of history is to be found in the period around 500 B.C., in the spiritual process that occurred between 800 and 200 B.C. It is there that we meet with the most deep cut dividing line in history. Man, as we know him today, came into being. For short we may style this the "Axial Period".
>
> From *The Origin and Goal of History* OGH

Commentary:

Jaspers' text indicates he is referring not to an 'age' so much as the transition to a new age: an interval from -800 to -200. The question is really about a larger process: a global context. But this view, applied not just to worlds in parallel, but to a sequence in time, provokes a 'dynamical' rather a theological or historical question: can a single turning point be taken as a 'starting point' inside a larger whole that started far earlier, primordially? It is clear the 'Axial Age' is so massively packed with innovations as to endorse Jaspers' conclusion. But it doesn't actually contain *everything* that fashioned humanity. The Neolithic 'Age' clearly fashioned a rival era, what to say of the earliest 'transition' to humanity, *homo sapiens*, sometimes in fact pegged as a definite 'axis point', the (controversial) idea of the Great Explosion, 'axial' age indeed, if real. We are confronted with an issue of 'relative' beginnings...

> Directionality and teleology are complex enough. Our data shows an exotic variant of an unexpected variant on unilinear direction. History shows splitting streams, as in our Axial Age. The contrast of 'stream and sequence' in our model can help to express the tension of directionality and parallelism, a strange minimax process on the way to globalization.

'macro' effect, the histories of 'god' concepts. Science attempts to banish 'god' from physics, but in the spirit of that proto-Kant, Plato, the 'edge of space' antinomy asks if we could reach beyond it. The antinomial character of basic metaphysical concepts was pointed out in the great endnotes to his famous *Critique of Pure Reason*, the so-called Dialectic. Between string theory and the edge of space we should pursue science, banish 'god' from physics, but acknowledge stalemate at the edge of space. The question of 'god in history' haunts the Axial Age, and we will consider the need to reconsider theistic historicism without reductionist illusions in their place. The current view attempts to posit the void in the emergence of something from nothing, and this is the atheist's version of the same creationist core that has been garbled by monotheism.

Fig. 6 Gustave Dore: *The Deluge*

Kant, despite a critique of design arguments spoke of a demiurge, the Sufi J. G. Bennett spoke of 'demiurgic powers' of nature, while Christians were very facile with beliefs about angelic realms, the 'Heavenly Host'.

This argument requires another of his ideas: the distinction of hyponomic, autonomic, and hypernomic domains, the zone of physics, biology, and a blank: the type of distinction arising in *Samkhya* of the 'spiritual' material, at the highest level. The existence of entities in the hypernomic zone with elements beyond life of consciousness and will are a logical possibility not considered by theists/atheists in their debates. The term 'consciousness', perhaps 'hyper-consciousness' might work better, is confusing and slippery.

We must be clear that while atheism satisfies the demands of many 'secularists' so-called, there is quite another issue in suspecting the 'existence' of 'spiritual' powers within nature, entities that exist, and thus are 'spiritually' material, in the vein of classic *Samkhya*. None of the antinomies of reason attend this hypothesis: 'god' must be beyond existence as its source, but powers within nature are excellent candidates for science fiction, on the way to being discovered.

Preface

Fig. 7 William Blake: Elohim creating Adam (note the distinction, 'god', 'Elohim')

The design debate, design vs. natural selection, is destined to be deadlocked. The strategy of reductionist scientism has failed here. But so has creationism trying to use design arguments as proofs of the existence of god. 'God' cannot 'exist' inside space-time, and is 'outside' of 'existence' in a different mode, being, beyond knowledge. This makes the whole debate nonsensical. And we cannot speak of 'intelligent' design in predicates for 'god'. Monotheists have lost the distinction between 'supernatural' and the 'spiritual' inside the realm of the material/natural, but it is present in the Old Testament as 'Elohim'. That then would be an empirical issue, a 'phenomenology' of 'spirits', the 'heavenly host' of the Christians, perhaps. Outlandish, but logical. In a sense the design argument should be a natural sideline to scientific research, since teleological machines are a staple of biochemistry, now confronting the epigenome. But this has nothing to do with theism, necessarily. The 'natural teleology' of Kant suggests that 'design' begins as a naturalistic phenomenon, whatever the mysteries of unknowable divinity. And there is a third possibility, as noted: a natural demiurge (plural?) acting within space-time, science fiction perhaps, but logical. The materiality of the 'spiritual' resolves the questions of material soul, short of the supernatural, which is beyond knowledge. The idea of a material soul (as opposed to an enlightened being beyond soul) is unknown to Christians, but is known in the Sufistic and Indic traditions. The problem with design arguments is, ironically, the way in which religious mythology has distorted the use of the term 'god', leaving it dangerously ambiguous, and design arguments fairy tales. The ancient prophets warned severely of the use of such terms of pop theism, reserving reference to a 'pointing to', as in the abstract referent IHVH. The strange record left by the Old Testament has actually lost the thread of its deep discovery of historical 'evolution', which can indeed impinge on design questions. But this record conceals a revolutionary discovery, which the creators of Israelitism did not yet understand. DMR

Because Bennett defined 'demiurgic power' in an occult way, we can propose a repeated change of terminology to expose dangerous thinking with an occult ambiguity: the idea is from Plato, the 'demiurge', in the singular. The plural is an apt speculation: we can however speak of the 'demiurgic atman/brahman' or the *demiurgic manifold* (or the manifold demiurge). In the conclusion we will use the term 'SPR-MAT-X', to create a neutral term. Note the similarity to the 'Elohim'. Modern skeptics can and should critique such notions, but they belong to

Fig. 8 Temple of Juno–Roberts

empiricism, of the future.

Design arguments next to the 'design inference', an idea created by the ID movement, but easily generalized, deserve this reserve hypothesis of the 'spiritual' inside nature, because it is an empirical question. Has mankind, as the Sufis secretly claimed, interacted with mysterious spiritual beings, upgrades of the angelic host? Skeptics must remain so, but might acknowledge this hypothesis can act as ground wire, and logically alternate to the hopeless confusion over issues of the 'existence' of god. We will not decide between theism and atheism, but will insist that the 'existence of god' is contradictory. 'God' would be beyond existence as its source in a larger realm of 'being', just as Escher diagrams are non-physical, but do have 'being'. Our historical model works fine without these Trojan horse concepts from fringe theology.

But the irony in the study of the Axial Age is the way

Fig. 9 Teertanker

Notes Toward an Evolutionary Psychology Beyond Scientism

Fig. 10, 11, 12
Woman, candle, bed, typewriter, violin

Schopenhauer spontaneously rediscovered the essence of the ancient Upanishadic/Samkhya spectrum of thought in the Kantian context, and he is the more convincing for having independently come on the core ideas, which degrade under repetition, and which began streaming into modernity in the Enlightenment. The philosopher J. G. Bennett recounts a variant of what Schopenhauer rediscovered, the ancient yogic-Buddhist distinction of 'being, function, will', in his language, of the causal order, its envelope of greater 'being', and the independent reality of 'will'. Despite its own problems (the 'will' is problematical as egoic or noumenal), this can help as an exercise, and to break the habit of the misleading 'material/spiritual' duality, which causes endless confusion. He cites the analogy of a wo/man inside a room with various objects, each with a function, a typewriter, a bed, a sewing-machine, a musical instrument, a telescope. In darkness the bed can be used, but not the machines. With a candle, the machinery can function better, while if window is opened, the telescope can be used. The machines correspond to functionality, the light to being, and the 'user' to the factor of will. Cf. J. G. Bennett, *The Dramatic Universe*, Vol I, (London: Hodder & Stoughton, 1953), p. 55. This analogy, where the light corresponds to 'being', and the will answers the question, who uses the machines? can help to sort out the confusion of brain, neuroscience and self, which has elements of being and will in a still unknown relationship to the functionality of 'brain'. Consider the distinction of physics and the equations of physics, to see that the discourse must embrace the larger framework of 'being' where the (Platonic) ideas of mathematics have we suppose a reality ('being') beyond physical existence. But Kant and Schopenhauer suggest the deeper insight into noumenal and phenomenal aspects and to the fact that self is larger than its categories of 'space and time'. Part of the problem of understanding thus arises from the way the 'mind' is embedded in a larger framework, which it can't see or visualize.

that it leaves us immersed in the study of modernity. The term 'secular' is being appropriated by the secular humanists who demand that modernity be atheist. But the early modern shows only a dialectic of resumed theism/atheism, after centuries of theocracy. The Axial Age is the source of the modern and shows a dramatic balance of opposites. Jaspers' views are those of a theologian of the German type: immersed in Kant and his successors. The combination is the only one that could have generated the Axial Age generalizations. We launch a preemptive strike against the use, misuse, failure of use, of 'designers' produced like rabbits out of a hat to answer to the intelligent design intuition given by our data:

> We can consider a set of 'spiritual' powers inside 'existence' at a very high level. These have been considered many times in history,

Fig. 13, 14 Moses and Akhenaton
As we move toward a new model of history we will discover a distinction of 'stream and sequence', the aspects of continuous and discontinuous histories. Moses and Akhenaton are in the stream aspect and not part of the Axial period.

with the onset of monotheism confusing the issue of their action. But this concept nexus is flawed and confused with the powers of polytheistic paganism and their sacrifices. We should consider only the abstract possibility of some functionality of higher nature with being and some aspect of 'will'. The action of such entities answers to the sense of design provoked by our data. But we must stay within the framework of our 'systems analysis' and not corrupt our data with designer myths.

The Axial Age enigma is resolved in a remarkable 'triple play':

> We are proceeding in several different directions, the historical, the evolutionary, and we find ourselves in the at first puzzling world of Kant. But his classic essay on history poses a set of questions about history and so does the thesis of the Axial Age: is there a connection? We are on the way to a remarkable triple play that solves the classic challenge posed by Kant, resolves the enigma of the Axial Age, and shows the relationship of both to our ideas of evolution and history.

There are three broad types of explanation for human emergence, evolutionary psychological, religious, and New Age. The religious myths have fallen by the way side, despite hints of ancient understandings in the Adamic corpus. The New Age/Indic accounts are highly suggestive. The category of 'evolution' is right, but not the darwinian version.

The attempts by evolutionary psychology to explain human evolution are almost more mythological than anything in religious or New Age speculative literature. This area is a void on all sides. But the evolutionary psychologists may well have insights into the microevolutionary adaptations of already existing 'man', often inducing decline of real potential. The suggestions in *Big Brain: The Origins and Future of Human Intelligence* (Lynch and Granger) are that at some point *homo sapiens* was more intelligent than he is now. The regime of adaptational natural selection is under strong suspicion as an eroder of human intelligence. The tale of Adam and Eve, charming even to the secularist, gives the game away, with a question: did man at the dawn of man interact with spiritual powers of some kind, and what did the Serpent mean in referring to consciousness. The psychology of evolution requires an answer to the meaning of the term 'consciousness' and the route by which its advanced potential emerged in man. The complex semantics of this term demands a 'fourth', in modern Buddhist 'slang', *enlightenment*. We must ask, if the San have a language with 120 phonemes, if man hasn't declined from his earliest nature, in which 'enlightenment' was his natural state, before the drift into mechanical states of mind. Beside the Buddhist there is the *Advaita* tradition which resolves the riddle in another way. Here the design argument lurks over the void of science: and the question of human evolution. The *Annunaki* are an internet obsession, but the clear mechanical character of the Axial Age shows us 'design mechanics' at its most elusive. But the Axial Age is a good portrait of the 'evolution' of religion. And a clue to the mysterious Old Testament.

The use of the formalism of evolution, macro and micro, is controversial, yet need not be, and is indicated by the data and greatly clarifies the blindness to levels of evolution. This is difficult for many, but it is a category nexus that exists in many fields. Does it apply to the data? Many will be critical in the name of science at the attempt to look at historical dynamics in this fashion. The reality is that 'science' fails with history and the use of a specified model as a trial of the data is almost the only option left. Here, the dramatic success, however limited, of such a simple model for such a complex phenomenon leaves us to wonder. We are onto something.

We will speak of a formalism of evolution: an intermittent driver operating on a system with two levels. Lamarck was the first to use such a distinction. We puzzle over theories, their status, proofs and implications. A 'model' is actually a better way: it is a trial against the data and a match allows us some measure of deductive insight. It corresponds to an idea of

Fig. 15 *View Into the Heyday of Greece*

'systems analysis', that is, a first encounter with a system too complicated for easy analysis.

> **From Jaspers:** It would seem that this axis of history is to be found in the period around 500 B.C., in the spiritual process that occurred between 800 and 200 B.C. It is there that we meet with the most deep cut dividing line in history. Man, as we know him today, came into being. For short we may style this the "Axial Period".

We need to move with this insight, and yet zoom out to see the limits of its validity: the real issue is that of a discontinuous turning point. The only way past the almost total confusion over the Axial Age is to see its place in a larger context of multiple civilizations. In fact, we deal in subsets of

Preface 19

Fig. 16 Shaman

Where does the shaman come from? The antiquity of this mysterious category of magician and 'spirit seer' accompanies any and all discussions of the 'evolution' of human consciousness with the companion issues of man's deeper consciousness, his will to the miraculous, and his own 'spirit' or 'soul', whatever that may be.

What is mysterious is the superstitious character of the historical remnants of this primordial category of man. The explanation is not far away: man may have had a far higher degree of self-consciousness in earlier eras of his emergence and the category of shaman, in some earlier variant, accompanied that passage with effects that were real. There is also the possibility of the now lost real meaning of 'incarnation' as the action of avatars, 'god men' (a term much abused, Bennett actually speaks of demiurgic powers entering early man), or men reaching states of enlightenment with active powers of the will. The rapid mechanization of descendants of such men provides a misleading history.

civilizations: 'macro' transitions, the key to the 'axial' riddle. It is useful to abstract the idea of evolution to a more general category that can apply to the historical. The evolution formalism is quite general and applies easily to the new context of world history. That is a direct challenge to the lame 'darwinism' crippled by its confused foundation. The 'macro' aspect of the Axial Age is a complete example. This formalism is a descriptive method, and merely points to a possible theory. The original symbolism arose to distinguish 'speciation' from the secondary process of adaptation. The levels help to distinguish emergence and realization. Monotheism shows an emergent character yet is a primitive human creation also.

Fig. 17 Ruins of Athens

I have given the book a second subtitle, suggesting the way the issues move forward to the modern world, provoking a discussion of secularism. The questions of religion, secularism and evolution, and, indeed, the evolution of religion, are a source of confusion in the context of the reign of darwinism. We are seduced into oversimplification in the name of science and the theory of natural selection has made us neglect the complexity of the question.

We need to face the reality that the data of the Axial Age explodes both secular and religious interpretations of the monotheistic myth of Israel. Further, the distinction of 'sacred' and 'secular' is misleading. The legacy of Old Testament history must face the challenge of a new Reformation

Fig. 18 Jacob and the Angel

to recast the histories of 'Israel' in a new global narrative. Our topics are the issues of the Greek, 'Israelite', Persian, Indian, and Chinese phases in a 'revised' 'Axial interval' from 900 BCE to 600 BCE. We will justify this change in the definition as we go along. But the ultimate strategy is to change

Preface

We should ground Bennett's idea of 'demiurgic powers', probably a variant of very ancient thinking, transmitted from greater antiquity by Sufism. A little known work, *The People of the Secret* (Octagon, 1983), by Ernst Scott, written in the wake of Bennett's *The Dramatic Universe* points to the connection with Sufism in that book, and also discusses Bennett's full thematic of 'demiurgic powers' and the 'hidden directorate', the latter being, evidently, historical beings in an esoteric realm generally unknown to the outer populations subjected to religious initiatives behind various forms of disinfo. This is an important distinction, since the confusion of esoteric human action and the far more elusive action of demiurgic powers operating on a far longer time-scale is a possible source of confusion. Much so-called esoteric action is on the level of barely competent confusion such as we see in the legacies of Rosicrucianism and the occult, while the realm of the demiurgic powers is on the scale of human evolution itself, scales not less than ten thousand years in basic time spans. The action of hidden occult powers can be very dangerous: we should never accept claims here uncritically.

If we stay within Bennett's analysis we should distinguish: 1. demiurgic powers, 2. an esoteric human sphere (hidden directorate), 3. the field of cosmic laws with the 'biospheres' in the cascade of the 'ray of creation'. The scale of the macrosequence shows a 'mechanical aspect' that might be biospheric, with the action of the demiurgic powers associated with that, like workers in a garden. This might help to sort out the hopeless confusion of the right interpretation of the Axial Age, and macrosequence. Note however that we disallow hybrid explanations. We have no proof of any designer power in history, and our systems analysis remains starkly silent about such entities.

The history of religion is beset with the claims of esotericists against religion, but these often fail to see the larger influence of the demiurgic realm which bypasses such influences in the larger direction of religion for humanity as a whole. Esoteric occultists are mostly deluded and have no direct spiritual authority. In general the religious sphere failed to predict or understand the rise of the modern and was taken by surprise by the onset of a new era. The chaotification of Axial Age religion in the result is clear. The mentor, Gurdjieff, of Bennett is one of the dangerous gnostics cited, and their disinfo has produced tremendous confusion.

terminology and leave behind the term 'Axial Age'. We can take a step backward here and use the term on the way to its larger context. Jaspers himself was on the way to this larger insight, but was unable to distinguish the issue of a 'civilization' from a smaller transitional interval. And the modern period confounds religious perceptions because its 'secularism' seems to be some kind of decline from a true spiritual age. We should disagree with but cite Jaspers' theological coloring of the 'Axial Age' but note carefully that it was a theologian steeped in German classical philosophy who was able to perceive the Axial period beyond its Occidental, Israelite, framework to posit a transcultural phenomenon. Here Nietzsche has confused many. He is a unique philosopher, but in reality anomalous. He is a 'come lately' far past the foundational moment of modernity, and he in no way produces the defining terms of the age of so-called secularism. He drives many to reject the modern world.

Fig. 20 Less known, Mahavir is a crucial Axial Age lead up for Buddhism

Jaspers hits on the extraordinary strangeness of world history in the way it shows set direction (his intuition about Christianity shows this) and yet multifurcates into multiple streams. The way to create a balance of direction and anti-direction is only visible in the larger 'macro' effect...We will approach this with a metaphor of stream and sequence...

Jaspers had breadth and was able to move past the Occidental, later Eurocentric focus, of the Christian world. Let us note that Jaspers' title to his classic implies a teleological interpretation. Jaspers might be confusing because he states one position and then moves on from it forthwith.

Fig. 19 Krishna scene Krishna is a mystery: not an Axial Age figure, but the outcome of a now unknown spiritual sourcing.

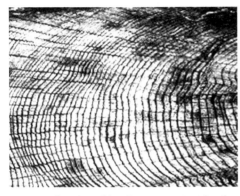

Tree Rings and Relative Beginnings.

We will have more to say about 'relative beginnings' but a simple example with illustrate the point: the layers of tree rings. It is perfectly possible to have an absolute discontinuous beginning inside a larger continuous stream, an invention is an example, but we should in this case be wary of the probable reality: a sudden beginning inside a larger stream is more likely one in a series of such. We should check the evidence here to see if this might be the case.

Each year in the growth of a tree is a relative beginning in a series indicating the age of the tree

This raises the issues promoted as Intelligent Design by religious critics of Darwinism. There is no real way to detect a designer. But the evidence seems overwhelming. What we can do is try to show the reality of design and then leave the evidence neutral. The result is a useful 'systems model' that tries to depict a complex 'design'. Consider the distinction of 'system action' and 'free action'. A shepherd and his sheep. Religionists must move beyond design, while scientists must confront complex systems that explain herding effects: input applied to systems that are aggregates of unit individuals. No accident the Bible is fond of sheperd metaphors.

Crucial is the 'stream and sequence' analog in the model. Splitting streams and directionality are a crucial aspect of the 'macro effect' and its model.

Fig. 21,2,3 Tree rings, sheperds, splitting stream

The idea of the Axial Age as the key turning point in human development won't work: a series of such transitions would make more sense. The origins of religion are, we muse, in the shaman and his mysterious world stretching back to the dawn of the Paleolithic. And Jaspers ignores the fact that the emergence of Christianity is not in the Axial Age.

The perspective developed here can help to see that historical archaeology and the new perspective of the Axial Age has actually discovered something arguably far more remarkable than the original. The drama of Revelation moves to become a chapter in a larger transformation of antiquity in the remarkable account of Jaspers, which is however caught between two stools. It is also important to see that the era of the Old Testament joins a family of such transformations, from Greece and Rome to India and China. In a final irony, if it is the case that the Axial Age shows the net equivalent of an 'age of revelation' on a larger transcultural stage, the same must be true of the modern 'next axial period', modernity. We should however refrain from that phrase to see that the *de facto* 'revelation' effect is real enough.

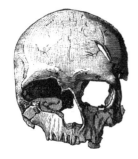

Fig. 24 Cro-Magnon man

Fig. 25 The First Artist: *homo sapiens* and the dawn of the artist

Our account will move in two directions and then unite both in a single spectacular solution to both Kant's Challenge, so-called, and the enigma of the Axial Age. The common solution to both problems is spectacular, and then impinges on a third, the meaning of 'evolution'. Our method is non-trivial and resolves one of the classic queries of the philosopher Kant in his essay on history. Our method of discovery is to detect a frequency pattern and adopt a finite-transition model for this to match 'directionality' with recursive cyclicity.

In *Chapter Two*, we present the solution to the riddle of the Axial Age,

Fig. 26 *God Bestowing a Soul on Adam*

The design debate has come to haunt Darwinian ideology. The strategy of reductionist scientism has failed here. But so have creationists trying to use design arguments as proofs of the existence of god. Monotheists have lost the distinction between 'supernatural' and the 'spiritual' inside the realm of the material/natural. In a sense the design argument should be a natural sideline to scientific research, since teleological machines are a staple of biochemistry, now confronting the epigenome. But this has nothing to do with theism, necessarily. The 'natural teleology' of Kant suggests that 'design' begins as a naturalistic phenomenon, whatever the mysteries of unknowable divinity. And there is a third possibility, as noted: a natural demiurge acting within space-time, science fiction perhaps, but logical. The materiality of the 'spiritual' resolves the questions of material soul, short of the supernatural, which is beyond knowledge. The idea of a material soul (as opposed to an enlightened being beyond soul) is unknown to Christians, but is known in the Sufistic tradition. The problem with design arguments is, ironically, the way in which Biblical mythology has completely wrecked the whole genre, rendering the use of the term 'god' dangerously ambiguous, and design arguments fairy tales. The ancient prophets warned severely of the use of such terms of pop theism, reserving reference to a 'pointing to', as in the abstract referent IHVH. The strange record left by the Old Testament has actually lost the thread of its deep discovery of historical 'evolution', which can indeed impinge on design questions. But this record conceals a revolutionary discovery, which the creators of Israelitism did not yet understand. If we confine these remarks to the *Preface* it could be because we won't solve them in the text! The issues are well cautioned by the philosopher Kant.

and this begins by citing a classic essay from Kant, on history. Kant called for a new Kepler to detect a natural purpose in the course of history. He projects the problem into the future, due to the need for future research. The nineteenth century amply resolves the need for data. It doesn't require a Kepler, a simple test of the data solves the problem very quickly, but only in a restricted range. Kant also refers to the idea of 'Nature's Secret Plan' and evidence for a progression toward the perfect civil constitution. Our analysis amply resolves this progression in what we have called the 'discrete freedom sequence'.

The revolution of archaeology in the nineteenth century produced for the first time in world history a continuous record at the level of five thousand years, the minimum as we will see for resolving the riddle of directionality. Two cycles and the start of a third will fit into such an interval, a tantalizing result.

Fig. 27 The Ghost Scene in Hamlet

The debate over evolution is confounded by the chronic uncertainty over the issue of design, now 'Intelligent Design' in the terms of a recent religious thinktank group. The issue of design is far too 'over-argued': the question of design is unavoidable. But by the same token the Old Testament has confused the issue of design in history, and we see design in the emergence of complementary religions, theistic and atheistic. The term 'design' is thus a very controversial one: but our account can help to see how religious and 'secular' accounts are both confused on the question. Let us note that one of the classic critics of design was the philosopher Kant. But it is also true that he suspected a thesis of a 'demiurge' to be at work in history. Beside Kant there is Schopenhauer with his insight into the 'Will in Nature', an ironic design argument from an atheist. Our model does its work without forcing the issue on the reader. So the 'hysteria'

Preface

Fig. 28 The Shiva Seal
Discussions of consciousness invoke questions about the origins of 'yoga' and the 'history of consciousness' in the Paleolithic and before. How much of the original human spectrum has died out?

In Bennett's scheme the material/spiritual is replaced with the hyponomic, autonomic, and hypernomic realms, the latter being the framework of four cosmic 'energies': conscious, creative, unitive and...? Man's animal vitality or awareness ascends an octave to 'consciousness' (self-remembering consciousness) which in turn is induced with a still higher 'energy' he calls 'creative'. To say that consciousness is a cosmic energy is borderline crackpot yet a cogent thought in the limits of his model. The tetrad shown is interesting but dubious. The point is that standard evolutionary scenarios, even those critical of darwinism, probably fail at this key point, the threshold where man as a hominid acquires a soul, a mind, creative powers, a sense of ethics, and most crucial a very complex linguistic reality with built in archetypes, myths, soon to become sagas, poetries, dance, etc... Bennett's remarkable portrait can hold the dyke of rampant new age speculation, keeping in mind that we have no real observations of the emergence of *homo sapiens*. It is an open question that man required direct avataric intervention to cross the threshold of man. A cosmic energy can clearly be detected by the 'vital apparatus' of man as he is, so it is a question rather of man attempting to overcome the limits of his animal being to realize his potential consciousness which as a cosmic energy is 'all around him'. Bennett's ideas are ripe for abuse, and one issue is the obsession in many legacies with 'sacrifice' and its place in the evolution of religion remains unclear. The world of dangerous gnostics Bennett lived in was obsessed with theistic class warfare and the 'food for god' rackets that emerged in the degeneration of now lost religious traditions where primordially the questions of animal/human sacrifice were born.

over design would be misplaced here.

Our strategy with design is to give it an airing but to focus on creating a neutral model as a foil to mediate discussion. Much of the critique of design is based on false assumptions, so we are forced to be wary of either side of the debate. It is useful to consider entities that are beyond existence, next to 'spiritual' beings that are within existence. The latter are the solution to the design riddle, perhaps. But this confounds the advance beyond polytheism that characterized monotheism.

The idea of a material/spiritual being within the realm of existence is ancient and has many forms, e.g. the heavenly host. The realization that such entities in various myths as bodies of light or the equivalent could exist helps to neutralize confusion over design arguments. Such ideas should remain definitional virtuality, and we must throw back a small fish as we discover our tendencies to recreate pagan fancies... they defeat the purpose of monotheism, very dangerously, yet show how the 'one god' idea was an idea of the 'demiurge' generalized past its Kantian boundaries, creating a legacy of confusion.

Angelic memes: As we see we are citing ancient archetypes (the Sufi versions are less known in the West). These are the same as our 'new' approach and yet we must renounce them at once: they are 'clutter' in religious history, which must recycle this concept ASAP due to metaphysical voiding of their status as monotheistic mythology (and apt substitutions for pagan divinities). Our solution is purely theoretical and posits material, living, and a third category: 'beings' with 'will' and 'consciousness' but not 'alive'. This is very strange at first but solves many problems.

Fig. 29 Ba Bird: enigma of Egyptian soul concepts

Dangerous gnostics The world of Bennett was an ambiguous interaction with Sufis and other shadow gnostics. We can only provisionally use his concept of demiurgic power: it will become the Trojan horse for all sorts of 'authoritarian explanations' based on esoteric deceptions. We can create our own version, and note that Christians had muddled versions of such ideas, such as the Heavenly Host. The point is that 'spiritual beings', fully material can exist within nature. Man is one of

The question of sacrifice in the history of religion is complex and hard to resolve. The Axial Age shows a clear initiative to move beyond the legacy of human/animal sacrifice. But this is delayed in the mysterious sluggishness and compromise of the Israelite tradition which is still in many ways in the realm of the earlier eras. The vestigial Abraham/Isaac meme has produced immense confusion and should never have survived into the corpus. The sublimation and abstraction of 'sacrifice' in the Christian legacy is also ambiguous, yet succeeds in leading religion beyond 'sacrifice' in the pagan sense. The confusions

over Abraham and the sacrifice of Isaac have been disastrous: is this not a hangover from the paganism of the earlier era? The confusion in India over the Aryan Vedism and sacrifice and the primordial Indian religion add to this murky history. Finally we touch on the question of the Maya and their synchrony…

Fig. 30, 31, 32, 33 Maya vessel with sacrificial scene, crucifixion as sacrificial sacrament, Greek animal sacrifice, Abraham and sacrifice of Isaac

them...This thinking is classic *Samkhya* and belongs to no one.

Soul hypotheses These are too controversial for current science, but arise the same way that 'spooky physics' arises, with the foundational conjecture that the human frame passes beyond the space-time boundary.

Bennett was kidnapped by Sufi pirates and used as a mouthpiece for some strange doctrines, but his insight into *Samkhya* is remarkable and belongs to greater antiquity and not to those who will take a proprietary interest in the use of his ideas: read only, consigned to the *Preface/Conclusion*: we cannot deal for long in these fictions, but they serve as temporary models to clarify theistic confusions.

The issue of secularism is very confused due to its shift in meaning. It is taken as a synonym for the rise of atheism in the modern world, but the term itself sources in the perception of the onset of a new era or epoch in the sixteenth century. It is thus ironically associated with the Reformation, not for the modern trend toward post-religion, a trend among many. A challenge confronts modernists: the crisis in the definition of the basic perspective we call 'modern'. Next to this lies that wild goose chase, the postmodern hallucination.

> The progression of thought, clearly visible in Jaspers, is to puzzle over the Axial Age, suspect the modern era shows a second such age, and then attempt to posit a set of axis points. This is the right strategy but it is not enough. The idea of a second Axial Age forces the question, has it already happened or must we create it? If the latter we confront the need to operate over many centuries to make changes in whole cultures. Not as yet realizable? The other possibility is that a second Axial Age has already happened. But religious proponents find this unacceptable: they wish to found a new 'Axial Age' to undo modernity which seems to lack spiritual potential. This confusion has haunted a very large contingent of thinkers including those in the phase of postmodernism. The problem is solved for us: the 'axis point' has already completed.

The question of religion in world history and in the modern world moving into a new future is very controversial because the reformation of religion is confused with its abolition via skeptical rationalism. But the influx of religious issues and cults via the New Age movement into the modern period in the nineteenth century shows perhaps how this issue will be decided. The question of epochs and ages is a study in itself, and the confusion over such thinking needs a correction. The issue is not the abolition of religion, but

Preface

Fig. 34, 5 Mayan priest smoking, Mayan temple

A puzzle inside a puzzle One of the mysteries of our analysis is the enigma of the Mayans: the Mayan start is in the Axial Age and shows the exact pattern of (Olmec) lead up 'stream and sequence' format that we see in all the other cases of the Axial Age.

This raises difficult questions for our model and for the issue of a civilization introducing 'sacrifice' in an isolated history that misses the coming abolition of sacrifice...We can defer on this confusion until the reader is clear on the nature of the dynamic involved. But we must exercise caution on these issues: rogue 'prophets', none other than Bennett's guru Gurdjieff being an example, trying to reinvent sacrifice for the future are analogous to union busting attacks on labor. They wish to roll back the gains of the Axial Age...

The problem has another obvious

solution: the Mayans show the stage of first state formation and are one cycle or more behind the average Eurasian field and are at a stage of development where sacrificial religion gets a pass from our macro effect. We cannot be sure at all of such interpretations, and we must be wary of any or all statements its Axial dynamics.

the recycling of religions of the Axial Age.

Our study of the Axial Age will bring home the issue of the evolution of religion in itself and in the context of modernity. We need an insight into the way new eras of history initiate and seem to produce transformations of religion, and not only religion but of the cultural totalities as a whole.

> Religion, along with advanced language, consciousness and 'soul' we suspect emerge with *homo sapiens* and his potential begins releasing in the eras of *Cro-Magnon* man/woman.

We need to see, as already noted, a dynamic that transcends the religious perspective. A closer look shows that the 'secular' era via the Reformation injected religion into modernity. Religionists dealing with legacy spiritual movements that emerged in the Axial Age would do well to study our model of history with its perception of epochs in a series. The process of renewal generates an imperative to religious critique that often disorients those committed to a traditionalist faith. The rise of modernity shows a consistent and repeated set of challenges to traditional religion. The issue is not that modernity transcends or discards 'religion' but that it will recycle the Axial Age brands.

This is obviously the result of the scientific revolution in its collision with archaic and non-evidentiary claims in the realm of theology, and history. This is taken as obvious by many 'secularists' but the questions of religion have not as yet been properly addressed by science. In fact, the question of what constitutes 'science' remains ambiguous in the realms of evolutionary theory, the study of human consciousness, and the nature of mind. The data of the Axial Age complicates the critique of the historical basis of Biblical Christianity and shows it to be a part of a larger Eurasian, thence global system. And the 'co-incidence' of the Greek Axial shows how the idea of an 'age of revelation' extends easily to what we might call the 'birth of secularism'.

The question of 'god' is intractable and it is of ironic note to consider that 'monotheism' has a dialect that refuses the term in the legacy of those who refused to utter the terms of divinity. The glyph IHVH, before it turns into 'Jehovah', can be put to this use for us here. We are confronted with a stern case of 'garbled message' syndrome: we have overwhelming 'evidence' for the action of 'god' in history, but a closer look shows that this isn't really 'god'. We need thus to be careful in assessing the claims for theistic historicism, mindful that the element of 'design' in history is altogether confusing and

Timeline for the Rise of Civilization

The factor of 'ten thousand year explosions' is clearly at work in the emergence ('evolution') of civilization.
From 50,000 years ago: dawn of human culture
20,000 to 15,000 years ago: the Last Glacial Maximum, transition to interglacial
15,000 to 12,500 years ago: Bølling-Allerød Interstadial, warming
12,500 years ago: Younger Dryas, 1300 years of renewed cold
11,500 years ago: onset of Holocene
Then around 9-10,000 BCE we see the first beginnings of the agricultural revolution with the Natufians in Western Asia. Then a new 'Great Explosion' takes place.:

The period of the Neolithic beginning ca. 8000 BCE in the Fertile Crescent is still too coarse-grained to detect the 'hidden transitions' we suspect, but we can plot the basic outline very easily:

-8000 to -5500 BCE is the first phase,
-5500 to -3000 BCE is the second phase,
leading to the take-off period of 'higher' civilization
ca. -3300 we see Sumer and Dynastic Egypt crystallize
A great field of civilizations and histories arise in the diffusion fields of these two great starts, but the basic framework is in place until the Axial period:
-900 to -600 BCE: we see a synchronous parallel emergence field across Eurasia, with ambiguous data for Africa and the New World. This massive convulsion of transformed culture sets the stage for the first stage of globalization.
1500 to 1800, another rapid transition to a new age period. The lack of synchrony here in the single focus of Europe is totally misunderstood. Our analysis suggests the obvious reason: parallel transitions would collide.
This almost miraculous pattern of data, alternating between fast advance and 'medieval' middles is a dead-ringer for the term 'punctuated equilibrium', and shows a clear frequency pattern of about 2400 years, as hard as that can be to accept. But this kind of action, totally unexpected, fulfills our requirements for an 'evolutionary' driver. Almost all the great advances of civilization occur within these 'axial' intervals. DMR

has been wrongly understood by all sides to theistic controversy.

The critics of religious historicism are themselves confounded by the perception of a process that exhibits the characteristics of the action of a 'higher power' in history. But the theological mythology that is traditional here is not adequate to this new data set.

The connection between the discussion of religion and the issues of evolution is a stumbling block for many. But the discovery of a large-scale pattern of the 'non-random' in world history forces the issue. Our use of the term 'evolution' will require a bit of caution. It will refer, not to a theory, but to empirical evidence of a 'developmental' process. And this evidence is present in world history.

We also have a way to connect

Fig. 36 Phoenician drachm, 4th century BC The coin shows a seated deity, labelled either "YHW" (Yahu) or "YHD" (Judea)
"A coin from Gaza in Southern Philista, fourth century BC, the period of the Jewish subjection to the last of the Persian kings, has the only known representation of this Hebrew deity.
The following from 'fringe literature on the Annunaki' shows the origins and complexity of the *Yahweh* term.
THE ORIGIN OF THE NAME OF GOD AND HIS TRUE IDENTITY.
Synopsis and Translation of the Phoenician, Ugaritic, Canaanite, Sumerian, Akkadian, and Assyrian Tablets. Kindle Edition by Maximillien de Lafayette

'evolution' and 'history' with a logical argument and this can help us to see that the emergence of *homo sapiens* is more than an anatomical discussion. The last generation has seen a remarkable resurgence of the 'design debate'. With man, the question arises as to how such a complex creature could evolve in isolation.

Remarkably biologists have boxed themselves into a corner via the claims that natural selection refutes the design argument. It has simply fueled the suspicion that design is a factor. The design argument is inevitable given the extensive and obvious plethora of designed biochemical machines dealt with by biologists. The larger question of design evokes the suspicion that it is an aspect of nature and that a teleological factor lurks in the whole game,

Preface

Theism/Atheism: The 'God' Debates :
Circular firing squad approach

The confusion of Darwin debate springs in part the attempt to use the evolutionary question as a battleground for beliefs in theism or atheism. Our brief discussion of Kant warns us of the intractable character of such debate, and the futility of this strategy on both sides. This polarization has become explicit in the crystallization of the so-called Intelligent Design movement next to the so-called New Atheists attempting, it seems, to make fundamentalist Darwinism a metaphysical foundationalism. In general, the context of the obsessive Western theism/atheism dialectic makes real evolutionary discourse almost impossible. The world has been held hostage to this closed debate long enough.

Richard Dawkins in his *The God Delusion*, along with Daniel Dennett in Breaking the Spell, have produced the symmetrical antithesis to the exploitation of the design argument in what comes close to claims for the legitimation of atheism in the assumptions of Darwinian natural selection. We can suggest that this is a mood, more than a philosophy, as the derailed freight-train of mechanized religion proceeds with dead momentum past all the implications of Enlightenment critique, threatening the attempted cultural renewal of modernity. But Darwinism is a poor candidate for meeting this trend. Religionists should take note of the inexorable dialectical reaction to stale theologisms in the ferocity of 'New Age' passages beyond the religions of antiquity, and the Axial Age. These 'New Atheists' are fighting the suffocation of stale theologisms.

The terms 'God, soul, mind, life, will, design, providence, consciousness, sacred, spiritual, transcendence' prowl like semantic wild beasts near any discussion of history.

We cannot arbitrarily exclude arguments by design, but we can demand new terminology, and precise definitions. We will make this our one inviolable rule. Thus, it is almost impossible to use the term 'god' without prejudice in relation to differing religions and our study will completely disallow it in any (theoretical) context. This is not an atheistic stance since the discussion is mostly meaningless, and it does allow fresh terms and definitions. Our position here is neither theistic, atheistic, or agnostic. These terms buttonhole all discussion.

even that the progressions of speciation have a hidden directionality. The question of design evokes the question of the designer, and this issue has foundered in theological distraction.

We should consider the 'existence' of beings of a high order who can be the source of unknown but highly advanced powers of action. This is a rubric of possibility that can free us from crude creationism, and answers the question of the 'who' in the designer category.

This issue arises as we attempt to understand human evolution. This is an intractable question with multiple question marks:

the emergence of language
the nature of mind
the question of ethical reasoning
the source of creativity
the enigma of 'self-awareness' as an 'instrument' of changing consciousness
the mystery of 'free will' or short of that, 'free agency'.
last, but not least, the issue of soul

Debates over free will can be metaphysical but our point is made with a lesser candidate: free agency. This is a staggering list of innovations, although their emergence in embryo at earlier stages of the hominid progression, especially that of *homo erectus*, provides some scope for evolution in stages. But over and over again biologists confront the necessity logically and the reality empirically of some kind of swift transition, already pegged in one rendition as the 'Great Explosion'. We cannot assume that such a complex combination of properties can arise piecemeal.

The problem here is that the darwinian framework cannot handle this 'almost impossible' set of interconnected innovations. The inescapable design argument, at first a mode of explanation for complexity, is transformed into a definite suspicion that the overall set of human characteristics had an external guidance process. It is actually more plausible that man's consciousness, rightly understood is a function of exterioriziing what is already there, so to speak. In any case, we do not understand our own consciousness, let alone is 'evolution'.

Let us consider that *homo sapiens* is a Buddha in embryo. How could this potential arrive by random evolution? Alfred Wallace, the parallel figure in the emergence of evolutionary theory, saw the problem of 'potential' traits, and moved toward a form of design argument. So we are in good company here. We confront the question of man's inability to pose a theory of evolution

Preface

Stream and Sequence

We can introduce a new and useful metaphor for the 'eonic effect', the 'stream and sequence' relationship. We can use this as another way of describing a series. Another related metaphor is a relay race, or pony express: a series of running streams in parallel, but the baton passes between different runners (streams). In the same way, we see a series of streams of culture, their long histories, but a set of short intervals promote a larger 'sequence'.

Stream and Sequence Consider the dynamics of the Greek or Israelite Axial intervals (or any other for that matter). A stream history leads up to the Axial interval and shows transformation. This transformation generates a higher level step in a greater eonic sequence. This is the 'stream and sequence' effect. We now have two levels to our account, the evolution of the stream of cultures, and the evolution of the high level sequence. And this allows us to give expression to ideas of evolutionary directionality and progress at the higher level. Or perhaps progression would be a better word. However, the idea of an eonic sequence allows us to proceed without committing ourselves on generalizations about progress which always end up confronted with various contradictions.

Consider the diagram below: the x's show sequence elements inside the continuous stream elements. A sequence moves between different streams...We have to wonder if an earlier Axial Age was not the case in the Neolithic, part 2...

```
1 xxxxxx_____
_____3 xxxxxxx_____
_____2 xxxxxxx_____
_____4 xxxxxxx_____
```

The pattern in world history is more complex and shows parallel effects:
```
xxxx----------------------------------------Neolithic, part 1
-----xxxx-----------------------------------proto-Sumer to the north, Neolithic part 2
----------xxxx------------------------------Sumer...3000 BCE
----------xxxx------------------------------Egypt

---------------xxxx----------------------Greece/Rome 900 to 600 BCE
---------------xxxx----------------------Isreal/Judah
---------------xxxx----------------------Persia
---------------xxxx----------------------India
---------------xxxx----------------------China
-------------------xxxx--------------------------Euro-sector: 1500 to 1800
```

that must explain complex states of consciousness that he has not usually experienced. This shows us that the issues of human evolution impinge on the issues of a kind of 'generalized' buddhism and this will undermine all simplistic forms of evolutionism.

The views of Schopenhauer suggest that an atheist can solve the problem in essence without the trappings of 'spiritual' metaphysics. The place of 'Will' in nature is one way to clarify much of the confusion created by reductionist science. Beyond this, yet part of the same set of issues, is the question of the 'will' in man: his intuitive egoic psychology of such, the stance of religious traditions on the question, the challenge of science to such 'superstitions' in the reign of universal cosmology. The questions raised by Bennett are thus a useful set of challenges to the usual understanding of human emergence.

WHEE was notable for its new model of history, and the basic elements will recur here, but on the sidelines: the simplest approach is to look at a progression of epochs. The reality is that world history shows a very remarkable dynamical structure.

The basic issues and concepts are:
the relationship of history and evolution: a deduction
the sequence of transitions and relative beginnings
a distinction between historical determination and free agents
a useful metaphor of 'stream and sequence', or directionality
a frontier effect that shows why the sequence jumps around....

The result, a portrait of directionality via a series of transitions, is called the 'eonic' effect, and/or the 'macro' effect.

There is a considerable literature online on the subject of the Axial Age, almost all of it confused. Best to shovel dirt over most of it and move on. Our model can help, use it on trial. There is a basic issue at the core of the confusion: science can't deal with synchronous events in parallel that have a common causal or generative core that must be in some sense 'trans-spatial' (but a 'field effect' works fine and isn't transpatial). And if the 'output' is diverse yet analog in parallel (e.g. two distinct religions), the problem seems to defy analysis. But our model is able to explain this situation fairly easily.

We can begin by citing the full text:
http://www.collegiumphaenomenologicum.org/wp-content/uploads/2010/06/Jaspers-The-Origin-and-Goal-of-History.pdf
Also: ascii format: http://www.columbia.edu/itc/religion/f2001/edit/

docs/axial_age.html

Much of the online literature on the Axial Age is of very poor quality. We could list some of the confusions here to suggest that something like our systematic 'systems analysis' can clarify a question that has antagonized most conventional historians.

http://en.wikipedia.org/wiki/Axial_Age
http://en.wikipedia.org/wiki/Talk%3AAxial_Age

Our approach here is too complicated for many readers. But a slow course of building up the concepts needed to study a novel and unexpected phenomenon with a stunningly elusive dynamics can help to clarify the issues, and to create a novel way of looking at both history and evolution. This approach is not dogmatic, and can allow reader to apply and then withdraw a model applied to something that defies intuition and which is hard to visualize. This approach will teach us, not to become believers in a new perhaps speculative interpretation, but to be wary of historical theories altogether. We can't get it straight by just reading 'continuous flow' histories. But the latter are much safer than attempts to reduce science to history and are advisedly the only safe way to approach the subject. Such works hit the tripwire of 'theory' is they dare dynamical concepts, like the 'decline of the Roman Empire', or the causes of the rise of capitalism.

Our method is 'idiot first systems analysis by the book', what is the phenomenon in question? World history (?what's that). Does the phenomenon in question show system properties: we can start first with the question, does the phenomenon in question respond to a frequency analysis of any kind? To our stunned surprise, the answer is 'yes' and we can detect a large scale system operating over intervals of around 2400 years. We don't need to pronounce this conclusively, because our data is insufficient to conclude the analysis. But the analysis, even if incomplete, can help us to forego the facile misperceptions of the Axial Age so-called.

The Axial Age and Its Consequences (ed. Robert Bellah and Hans Joas)
http://www.hup.harvard.edu/catalog.php?isbn=9780674066496

Religion in Human Evolution
From the Paleolithic to the Axial Age
http://www.hup.harvard.edu/catalog.php?isbn=9780674061439

The first work from many scholars contains a lot of doubtful analysis, and some drastic errors: who on earth can decipher the history of Indian religion? The necessity of making the Axial Age safe for Darwinism results in the reiteration of all the fallacies of evolutionary psychology.

http://muse.jhu.edu/books/9781438401942
The Origins and Diversity of Axial Age Civilizations
Shmuel N. Eisenstadt (1986)
This work has a large number of useful observations but the analysis cannot get out of the confusion in treating parallel events as having a significant relation while treating them as subject to a single historical stream. But the author begins to realize that Axial Age Greece is a part of the analysis.

http://books.google.com/books/about/The_Great_Transformation.html
Karen Armstrong's *The Great Transformation* shares the common fault of confusing the Axial Age entirely with religion, and then trying to find the common denominator of Buddhism and monotheism, which won't work.

http://www.baylorpress.com/Book/396/Convenient_Myths.html
Convenient Myths
The Axial Age, Dark Green Religion, and the World That Never Was
By Iain Provan
This work actually produces a challenge to our work, a recent interpretation that includes Suzuki's discussions. But we are free of the myth of the Axial Age: it has no final status in our fuller model! The problem here is the way the treatments of the Axial Age as stated are easily shown to be dubious. Our usage is *en passant*, and ends with a larger model. Is this model a myth? The advantage of a model is that it is proposed for examination, and is not stated as a final conclusion. And this particular model folds its cards on the Axial Age and looks to a large pattern. Jaspers actually senses this and his second paragraph rescues his book from the fate of most with its correct shift to the term 'axis': an axis of history would have to be empirically given. This approach survives handily into our larger model with its sequence of 'axis' points.

Preface

The thinking of J. G. Bennett, which springs form the legacy of Sufism, might help (to make us realize our ignorance). He speaks of demiurgic powers, of biospheric action, and almost of another category, the realm of what he calls a 'hidden directorate', evidently living men who have attained some higher state of spiritual consciousness, able to act in relation to spiritual or religious categories. It is also a question of 'soul' but in a Sufi sense. We have to be suspicious that something like this is present behind the scenes in the emergence of a religion like Christianity. But the category of *hidden directorate* is very dubious, and figures like Bennett are suggestible disciples. Figures trained on the spot via spiritual awakenings are the only likely candidates. A hidden esoteric group is going to be a dangerous mafia. We must be wary of such concepts. The 'hidden directorate' was taken by surprise by the rise of modernity, a failed test so basic we must suspect the status of the idea is dubious. But Bennett's scheme, with our modification, makes sense. Perhaps he was right. Clearly the culture of Israelitism in the centuries after the Axial Age had a potential (there are even dubious predictive myths pointing to the future realization of such a potential) but this could only be realized by some power that would have to be distinct from the Axial system which had completed its action. There is something mysterious in the triadic appearance of three men, John the Baptist, Paul the Apostle, and the figure 'Jesus', not yet the Christ. This strange drama is almost unfathomable and rapidly generated so much unwitting disinfo or instant mythology that it is hard to make sense of it. But clearly we have an instance of the category of a 'hidden directorate'. And the myth even includes intimations of this in the strange tale of three Magi, who knew what was afoot. Why would this be different from a putative category of *demiurgic powers*? Isn't our explanation getting out of hand with a multiplication of entities? The charge could be just, but in fact the extra detail might start to make sense of what never made sense, the New Testament. No other approach can explain how a religion with such a harebrained outcome as the Christian could also have a high spiritual source, or sources. The triad at the start remains a clue/riddle we can't solve. The answer is clear in fact: some group of spiritual beings in the background helped to jumpstart Christianity with a triad of founders, and this appears most remarkably to have occurred in concert with the Mahayana phase of 'savior religions' trying to transcend its own legacy of 'Buddha religions'. Is there a difference? Anyone who tries to 'enlighten' a whole culture thence a whole species will soon repent of such an ambition, and move to rescue the project from disaster with a savior religion, a life boat wherein the intractable effort to get a sutra in the hands of the multitude will take the form of a 'bible' preaching 'salvation', safe passage of beings to a distant shore. To save people is thus a more basic task than the generation of completed buddhas.

Note: an advance conclusion in terms of Samkhya

We will consider one form of conclusion before starting, with in addition an appendix, the last section of Chapter Two, and the *Conclusion* proper. This constant repetition may be useful, and the stages of the conception emerging here. It uses an idea of *Samkhya*, and a version of the Sufi ideas of Bennett, and the distinction of demiurgic powers and a hidden directorate.

> Our original model works fine, but leaves a mystery, but the introduction of these extras ideas can lead to confusion. We can dismiss them and return to the original neutral model.

We are really dealing with a clever update of the ancient *Samkhya* and its 'Sufi' signature is an appropriation. *Samkhya* has decayed into a 'dualism' but its real point is a version of universal materialism. It actually mimics its ancient antagonist, *Advaita*, in making 'consciousness' stand outside of the whole matrix. That only makes sense in the explanations of its enemies, the *Advaitists*. The use of the term 'consciousness' in Bennett might have benefited from this other Indic lore. Both perspectives might be true: consciousness is an evolutionary emergent in the animal to man spectrum, and also a 'universal category' of non-dual core reality, whatever that means.

We can also consider Bennett's distinctions of *demiurgic powers* versus a 'Hidden Directorate'. We will add the 'Biosphere' to this. The later is real, but not in the usual often claimed. The following is so complicated it almost explains why the religious mythology defaulted to a narrative of 'god' in history.

The 'Hidden Directorate' is a much hyped category of spiritual humanity as a 'gang that couldn't shoot straight', while the demiurgic powers have been corrupted by Bennett and elevated into a gang of cosmic cannibals in the legacy of the dark Sufism of the gnostic groups Bennett frequented. The idea that mankind is directed by hidden mystics is a confusion of history. Our model of history shows the reality. But we can see his point: the sudden appearance of three figures like John the Baptist, Jesus, and Paul the Apostle shows the action of men within history, but with connections to a spiritual legacy. This is NOT the explanation of the Axial Age, please note, which is far more difficult and has nothing to do with the mideonic appearance of Christianity or Islam. The Axial Age must find an explanation that is a

Preface

A 'dialectical' self-contradiction

Although our trend toward systematics attempts to rule out theistic interpretations, we should close with a kind of neutrality on the 'god' question. A systems argument is one step backward from a causal argument and is thus a description, not an explanation. We cannot therefore say we have reached an 'atheist' conclusion. Modern atheism is an attempt to escape the near labyrinth of confusions created by 'pop theism' and the failure to honor the IHVH glyph with silence. Our model enforces that silence, even as it exposes a complex 'design' which appears to show the action of the 'will' in nature, to use a phrase of Schopenhauer, the mirror image of the theist. His argument is open to the equal challenge of the metaphysical posit of knowledge of the noumenal. Here we come closer to an understanding of the IHVH glyph. Let us cite a true act of religious skullduggery: taking Schopenhauer's model and replacing 'will in nature' with 'god'. The reverse is also possible. Such a transposition explains a lot about the still mysterious birth of 'monotheism'. If we see how fast our 'transitions' decay we should wonder indeed what has been lost.

We have cited the classic antinomial metaphor of Plato: the idea of the 'edge of space' and the attempt to reach beyond that 'limit'. We confront one of the earliest recorded antinomies of the type catalogued by Kant at the conclusion to his *Critique of Pure Reason*. Both rational theology and rational physics are confronted with this realm of Red Queen's summary judgment of final theories. We can see that as in the Koran's reference to 'that' which is as close as the jugular vein, we must concede as rational physicists that the antinomial, if not hyperspace, is 'omnipresent' at all points. It is simply 'faith' taken in two ways to claim that 'that' cannot penetrate at all points with 'acts of will', that is 'god', or as 'nothing' with a creationist nothing that produces everything from nothing.

The Axial Age as an example of synchronous 'everything from nothing' inside an earthly everything on a scale so spectacular we should forgive the 'explanation' of IHVH downshifting into Jehovah. As to Jaspers claims the Incarnation was the center of history, we should say, wait and see, and note the 'origin and goal' of history is a universal cosmology yet to be written.

common denominator for Axial Greek 'eonic emergents', Israelitism and its Persian companion, Indic religion, and Chinese Confucianism and Taoism, a problem of such staggering complexity we are left aghast.

It is almost impossible to explain how a religion as confused as Christianity could be the result of a higher spiritual power. But our model deals with this fairly easily. In the transitions, system action and free agency as distinct yet unified shows how primitiveness persists. In the mideonic periods, free agency contrasts with the action of a spiritual power. A Charles Chaplin movie can result!

> We can simply redefine 'demiurgic powers' as being in the realm of *Samkhya's* 12 gunas, without physical bodies, but with 'will' in the realm of nature, with or without a distinction of noumenal and phenomenal. The question of the biosphere is unknown to us, but in the *Samkhya* scheme of Bennett, its causality is in the realm of the 'will'. Cosmic entities have 'will', but in a special sense. The real solution to the Axial mystery is at the level of the 'biosphere'.
>
> Note the strong resemblance to Schopenhauer. Bennett ended up using that philosopher to expound the otherwise muddled legacy of *Samkhya*.
>
> We must pursue the question of forms of temporal being that have 'will' but not bodies in the life realm. If we allow such a possibility we can resolve the design question in the context of a spiritual realm beyond life but within existence as a subtle and still unknown form of spirituality. Many problems disappear with this set of hypotheses. We can leave these as speculations, but the process of redefinition may itself be sufficient to relieve our confusions.

Our survey of the Axial Age confronts us with a stunning mystery. This spectacle of the Axial Age has been interpreted too often in isolation, but the evidence points to its inclusion in a larger pattern of emergence. Because of the way Jaspers posed the issue of this mysterious pattern of data, it has been taken by many as a generalization of the idea of an 'Age of Revelation', but the real implications are much different. The emphasis on sages and prophets is misleading and Axial Greece and Israel show clearly that geographical zones of parallel cultures undergo integrated transformations as a whole. That is something far more complex and demands a new kind of model to try and get a fix on what is happening.

Our data resolves one set of questions only to raise a whole series of different ones. We have created a juxtaposition of the whole data set with that of the rise of the early modern which, by our analysis, is a kind of 'next Axial Age'. This is very troubling for many who confront the many confused declarations about secularism. As we have pointed out the secular is not an antithesis to the religious and we see the modern Reformation as a clear indication of this. In the rapid diffusion of global religion in the late stage 'new age' movement, we see that, if anything, modernity has amplified the effect of the religions of antiquity. But the core early modern shows its true colors in its rapid transition beyond Axial Age religion to a new world of science, philosophy, politics, and the 'secular' phasing of religion, which means that it is has broken the theocratic grip of medieval Catholicism.

The idea of a 'reformation' immediately provides a clue to the Axial phase of ancient Israel as a 'reformation' of the type of polytheistic paganism present in the broader Canaanite and middle eastern milieu of its arising. Its challenge to idolatry with the idea of the 'one god' is so classic and so appropriate in its context that we forget than the now lost hints of an austere conception of the 'unutterable name of god' as IHVH that stood as the source of what soon became the 'pop theism' of a Jehovah god, one immediately dressed in the new mythology of the Old Testament. It should not surprise us that the rise of modern atheism has moved to expose this as the last stand of idolatry!

We have replaced the idea of an 'Axial Age' with the idea of an 'interval' of transition leading to a new age or epoch of history, and in the wake of the axial interval, whose chronology we have revised to three centuries from 900 BCE to 600 BCE, we see a new era that will spawn several world religions. Note that the religions of Christianity and Islam occur after the Axial period in the 'mideonic' or middle period of the new epoch. Thus their character is different, and this leaves us with two mysteries. Our model immediately exposes this and many other elements that tend to confuse understanding. The whole question of these religions is 'eonic': they show sources in a transition phase, but a realization much later in the middle period. This pattern most remarkably is isomorphic with the instance of Buddhism in India. And there again, and in remarkable concert with Christianity, Buddhism spawns a 'savior religion' variant of its own Axial source cult. That is powerful evidence that our 'system analysis' is on the right track in trying to find something more abstract than the action of a divinity behind the emergence of a theistic–and an atheistic religion.

In fact the synchronous case of Axial Greece shows us the most general case which is not even focused on religion. In fact, there is a religious component to the Greek case: a kind of last flowering of aesthetic polytheism, next to a clear 'first birth' of modern 'secularism' in the explosion of innovations in art, philosophy, politics, science, and many other categories. The case of China must accompany this analysis, but with an ambiguity that is appropriate to our larger study.

We can consider the fence-straddling double aspect of Confucianism as a hybrid religious/secular Axial Age phenomenon. A work such as *Confucianism as a World Religion* clearly makes the case for a religion here to accompany the parallel monotheism and dharmic Buddhism of the Axial period. But Confucianism (with a matched Taoist legacy) is also another proto-modern or 'secular' premonition. Overall the whole of the axial era has an aspect of rationalization of the legacy traditions, and we see this counter-intuitively in the assault on polytheism.[2]

The mystery of the Axial period is difficult to resolve. We have distinguished the action of the 'axial interval' proper and the mideonic realizations that come later. We attempted to create a potential field of explanation in pointing to a biospheric or '*Gaian*' level of action in the clear global phenomena of the Axial period, and the more general macrosequence. Next to this we have the mideonic creation of religions (and this includes the creation of 'Judaism' in concert with Christianity) as such and this is a complex hybrid of the sources arriving from the 'axis point' and the action of men, in concert with unknown spiritual powers of a different type.

If our model does nothing else it points to such at first obscure distinctions that other types of historical sociology are blind to. But this leaves us with many questions, and with Christianity we have to ask, how did this religion arise and who or what was behind it? Our model is very abstract and allows only two modes, *system action* and *free action*. System action is in the category of 'formal evolution (macro)', non-darwinian, with free action points to some form of free agency. It corresponds to the 'transition' and its action seems correlated with the biospheric level as a 'system', but connected with 'will' in the sense of Schopenhauer/*Samkhya*. It seems to transmit a 'Goldilocks principle'.

The factor of free action can be the action of ordinary men, or, 'for all we care', the ghosts of dead Egyptian gnostics. It can also be that of demiurgic powers. Historical sociologists of the Weberian cult will be appalled at such a statement, and that is not inappropriate, but such an outlandish suggestion merely points to our inability to answer the question, how did a religion

---

2 Anna Sun, *Confucianism as a World Religion*, Princeton, 2013

such as Christianity arise (its echoes of Egyptian gnostic themes has been repeatedly pointed to, the reason for our suggestion)? Our problem is of course still more complex because Christianity is a hybrid with Zoroastrianism, while the latter will generate ultimately the parallel mystery of Islam. Our model thus gets the answer, but it is clearly incomplete. The answer in one is obvious: it is a crystallization of Axial Age themes. The point to understand is that once the 'system action' of the Axial interval is complete, something comes into existence that is open to free action. And this might realize the source potential, e.g. the creation of a 'universal religion'. Our remarks on the 'Hidden Directorate' can help here.

> Clearly the culture of Israelitism in the centuries after the Axial Age had a potential (there are even dubious predictive myths pointing to the future realization of such a potential) but this could only be realized by some power that would have to be distinct from the Axial system which had completed its action. There is something mysterious in the triadic appearance of three men, John the Baptist, Paul the Apostle, and the figure 'Jesus', not yet the Christ. This strange drama is almost unfathomable and rapidly generated so much unwitting disinfo or instant mythology that it is hard to make sense of it. It is another gnostic riddle. But clearly we have an instance of the category of a 'Hidden Directorate'. And the myth even includes intimations of this in the strange tale of three Magi, who knew what was afoot. Why would this be different from a putative category of demiurgic powers? Isn't our explanation getting out of hand with a multiplication of entities? The charge could be just, but in fact the extra detail might start to make sense of what never made sense, the New Testament. No other approach can explain how a religion with such a harebrained outcome as the Christian could also have a high spiritual source, or sources.

> The answer is clear in fact: some group of spiritual beings in the background helped to jumpstart Christianity, with echoes of Egyptian gnosticism, with a triad of founders, and this appears most remarkably to have occurred in concert with the Mahayana phase of 'savior religions' trying to transcend its own legacy of 'Buddha religions'. Is there a difference? Anyway who tries to 'enlighten' a whole culture thence a whole species will soon repent of such an ambition, and move to rescue the project from disaster with a savior religion, a life boat wherein the intractable effort to get a sutra in the hands of the multitude will take the form of a 'bible' preaching 'salvation', safe passage of beings to a distant shore. To save people is thus a more basic task than the generation of completed Buddhas.

The 'Hidden Directorate' is a case of Bennett's suggestibility by his gnostic friends. The Gurdjieff question is closer to 'Hannibal Lector' occult melodrama than serious spirituality, so we should be clear that this material from Bennett refers to a late phase of his work that was influenced by spiritual powers or persons not connected with Gurdjieff. He met a large number of Sufis over the years, so it is difficult to understand his sources. The treatment of modernity (very poor nonetheless) and the sudden new age demarcation of the 1848 period show a mysterious influence of an 'esoteric left' attempting to counter the reactionary bias of that rogue Sufi world. We see that he had already left the Sufi world, and was in contact with something radical. His *The Dramatic Universe* should be taken critically in any case.

The intersection of Buddhism and the Occidental monotheisms is remarkable evidence of the deeper complexity of the Axial Age and its succession, while the place of Buddhism in the Indian Axial Age is still another, one we must defer to another book perhaps. But the buddhist legacy often confuses holders of the ancient Indian legacy (which degenerated into the 'Hinduism' we see today) because they could not see what we see now, the way in which a strain of the Indian religious legacy was used to generate a global version of itself as a universal religion. This streamlined and yet potent version of the tradition was a revolutionary construct that both echoes its ancient past and yet prefigured a future 'secular' era, and we the say its remnants have already slipped into the modern world aspects of the modern 'reformation'.

INTRODUCTION

1.1 The Enigma of the Axial Age

The study of world history has been revolutionized by the remarkable discovery of the so-called Axial Age, a term given to the data by the philosopher and theologian Karl Jaspers. This discovery is the result of the explosive growth of archaeology and historical research in the nineteenth century. A fitting starting point might be the work of Champollion, although strictly speaking the question of archaeology is as old as civilization and certainly shows a beginning of sorts in the wonder shown by Egyptians of later dynasties, confronted with their own history in the spectacle of monuments in sand.

The data of the Axial Age is 'déjà vu' all over again, in the sense that it was first observed by the redactors of the Old Testament, who were the first men in world history to record an 'axial age transition', in the world historical first, the Old Testament. Theological disputes can blind us to the incipient scientific character of the effort, albeit once cast in the sage of Canaanite religion. That religion under transformation, recorded by its immediate successors, is one of the wonders of world history, even for a secular age beset with its own quite 'axial' question.

The discovery of the Axial Age is a fitting accompaniment to the Age of Globalization, for it shows that global civilization is the result of something almost mysterious, and it is not a modern innovation: we see in the period in question a spectacular display of parallel synchronous emergentism, across the entire field of the Eurasian continent. This set of effects from Africa to China, with a question mark about the New World, is evidence of something acting with almost *Gaian* force, at the level of a planetary species. And that raises the question of the evolutionary as the suspected context for this prodigy of greater nature.

Fig. 2 Heraclitus

Fig. 1 Temple of Zeus: the 'Axial Age' in a larger model becomes a fixed frequency output that works on what is in its mainline. Thus Axial Age Greece shows a final flowering of polytheism

Fig 3 Isaiah

The question of the Axial Age poses a riddle for us, as we look backward at world history through the lenses of modernity, which itself poses a question about discontinuities in history. The Axial period provokes a crisis of theory, due to this sudden break in the record of continuous histories. We are forced

Introduction

The Axial Age

Karl Jaspers called the global transformation from -800 to -200 BCE the 'Axial Age', and its effects are visible in Greece/Rome, Israel/Persia, India, China, with probable effects in the New World and sub-Saharan Africa

The phenomenon of the Axial Age shows synchronous effects across Eurasia

The Evidence of World History Our increasing knowledge of world history answers Fisher's lament. Ironically, it is also world history that can assist us in answering our questions about evolution, the evolution of man. We can see that the issue, for example, of facts and values is intrinsic to development. This fact alone should alert us to limits of reductionist accounts. But there is more, a surprise: world history is actually beginning to show us a mysterious dynamic behind its seemingly random chronicle.

An Empirical Breakthrough: The Axial Age One aspect of our transformed view of world history is the discovery of the data of the so-called Axial Age. The question of evolution has been confounded by this discovery of a massive non-random process at work in world history. This discontinuous global process gives evidence of a dynamics of history that we had not suspected, and which throws light on the history recorded in the Old Testament. The Axial period shows us

1. a clear example of the way discontinuity can arise in a temporal historical stream,

2. how synchronous emergence can occur in a parallel, multitasking set of processes,

3. that there is global aspect to historical 'evolution', contradicting the standard insistence on local micro process,

4. purely cultural transformations are central, beyond the assumptions about purely genetic change.

into the camp of the ancient Israelites who clocked a great transformation and rendered that in the fantastic images of Canaanite epic, there to demand the basics of a science that does not yet exist, the science of history, if it can be called that.

In an age beginning to explore the extra-planetary dimension of cosmic space, the perception of the Axial Age is a fitting starting point for thinking in terms of a planet, and, indeed, of a species, that of *homo sapiens*. The nature of history forces us to ask for its dynamical laws, only to find the stubborn exception of the factor of willed agency, and this in turn begins to trespass on the question of evolution itself. It is here perhaps that the discovery of the Axial Age can suggest a solution to the paradox of historical emergence in the context of the speciation of that remarkable hominid, man. And in the process the question of the evolution of religion comes to the fore with a suggestion given by history that the stance of the ancient Israelites contained a first glimpse of the answer.

Fig. 4 Ascetic Bodhisatta Gotama with the Group of Five.

The rise of modernity has confounded religious histories, and produced an often misunderstood brand of 'secularism'. We are confronted by the epochal transition of the religions of tradition, as they are challenged by the rationality of the Enlightenment. But the theme of the secular modern often forgets the place of the Reformation in that passage, and that period of 're-forming' contains its own hint to our enigma.

Fig. 5 Confucius

We must go in search of history to find the significance, and connection between the secular, the religious, and the evolutionary.

1.2 A Second Axial Age? The First and the ...Seventh...

One of the steps in understanding the Axial Age is to see the seeming recurrence of an 'axial age' in the rise of modernity. The problem is that the

Introduction

> **An Axis of History?**
>
> An axis of history, if such a thing exists, would have to be discovered empirically, as a fact capable of being accepted as such by all men, Christians included. This axis would be situated at the point in history which gave birth to everything which, since then, man has been able to be, the point most overwhelmingly fruitful in fashioning humanity; its character would have to be, if not empirically cogent and evident, yet so convincing to empirical insight as to give rise to a common frame of historical self-comprehension for all peoples–for the West, for Asia, and for all men on earth, without regard to particular articles of faith. It would seem that this axis of history is to be found in the period around 500 B.C., in the spiritual process that occurred between 800 and 200 B.C. It is there that we meet with the most deepcut dividing line in history. Man, as we know him today, came into being. For short we may style this the 'Axial Period'.
> From *The Origin and Goal of History*

We see the genesis of the idea of the Axial Age in Jaspers, still halfway between a religious and sociological concept or model. Basically the legends of an Age of Revelation self-generalize to a perception of synchronous emergence and a sequence of 'Axial' periods. Jaspers overstates the uniqueness of the Axial period, and we can see the more plausible extension of the idea to a series or sequence.

phenomenon of the Axial Age finally makes no sense in isolation.

> Jumping to our result: our strategy is to see if there is any frequency interval that can link periods before and after the Axial Age as a series... The answer is found by trial and error: 2400 hundred year intervals link 3000 BCE, 600 BCE, and 1800, in each case with a lead up phase of what we can call transitions.... We can also try to move further backwards...
>
> ?Interval 1   10k BCE...Natufian?
> ?Interval 2   ?Proximate start of Neolithic ca. -8000
> ?Interval 3   ?The Middle Neolithic interval ca. -5400
> Interval 4:   The era of Egypt, Sumer, interval before -3000
> Interval 5:   The 'Axial' period, interval before -600
> Interval 6:   The early modern, interval before 1800
>
> In one stroke the whole of world history shows a coherent sequence, but we must be wary of any interval where the data is still very thin, before 3000 BCE. Note the importance of 'relative beginnings', e.g. interval 4.

Looking at this Axial phenomenon we are confronted with an inexplicable mystery. But one clue to the riddle lies in seeing that this period is not unique, but one in a series. The resolution of the mystery comes to us quickly, as long as we are not distracted by the interpretations of the Axial period solely as a spiritual age of religions. We ask, are there any other periods like this? The great clue is the remarkable resemblance of the Greek Axial interval and the sudden rise of modernity from 1500 to 1800. Moving in the opposite direction, can we find a similar period of rapid innovation and sudden advance? We don't have far to look. We suddenly see that the birth of civilization, and the rise of modernity are different phases of a larger pattern, with the Axial Age in the middle. Seeing the rise of the modern as a kind of second Axial Age suddenly makes sense of the data. In fact it is a third, at least, the extraordinary rise of Dynastic Egypt and early Sumer being a giveaway. We are forced to consider that the Axial Age is really a step in a sequence, and moving backwards and forwards we suddenly discover the full pattern. We can see three turning points equally spaced, with an interval of about 2400 years, clear evidence of a cyclical phenomenon.

The question of the Axial Age has spawned a new historical myth of a spiritual age producing the world's great religions. The fact that Buddhism (and Jainism) are 'atheistic' while the Israelite Axial interval spawns a theistic religion makes any simple interpretation highly problematic. The case of

Introduction 55

Karen Armstrong in this YouTube video discusses the idea of a Second Axial Age, We are confronted with a new ambiguity: has a 'second' Axial period already occurred or should we be thinking about how to bring it about. This provokes the crucial question of free agency versus some dynamical argument about age periods, which apparently ought to be on a larger scale than volitional action.

As if the problem weren't confusing enough we are confronted with the problem that after all this talk about religion, the genesis of Christianity, for example, is not in the Axial Age interval. Can't we found religions freely at any time?

Armstrong again is reflecting on the nature of modernity, and the need for a resolution of the religion/secularism question. Somehow secularism seems odd: it isn't expressing 'religious' values.

https://www.youtube.com/watch?v=uKi7bB4XgxI

Our discussion of Kant brought out the question of freedom, free will, free agency and their relation to causality. This distinction allows us to speak of free agents in relation to some kind of 'determination' overall. How do we reconcile this paradox? Also, were the creators of the Axial Age free agents?

A useful exercise is to consider a free agent before, during, and after an 'Axial Age'. We could not create a successor to such a period due to its macro determination. Nor could we initiate such a period at any given point in history. We can observe such periods only after they have occurred...

Greece is then downplayed because it doesn't fit the religious pattern (it actually shows a last great flowering of polytheism along with the seminal emergence of a critique of such). The pattern is far more complex than an association with transcendental mythologies. If there were ever an age of 'revelation' it has to be the Greek case, whose multidimensionality is spectacular. Out of the blue, a frontier area relative to the Middle East undergoes a prodigious flowering. Note the extraordinary synchrony of the core Old Testament period of the Prophets, and Archaic Greece. Then note how the Indic zone recycles itself in Buddhism, Jainism, etc, stripped of all local associations with 'Hinduism' (a highly vexed term). In fact, 'Hinduism' itself recrystallizes as almost a new religion. Our historical dynamic thus transcends the content enclosed in the remarkable 'Axial interval'.

Fig. 6 Cave painting

The problem is the extraordinary parallelism that places the 'Axial' period beyond anything to do with religion. This is also the era of the birth of democracy, science, and the proto-secularism of the modern period. These are all pups from the same litter in what must obviously be a form of multitasking parallel evolution, a shotgun effect exploring different possibilities. The Axial Age appears at first to be unique, but then shows itself as a step in a more general pattern, perhaps a sequence? With this question the real antecedent and continuation suggest themselves, the birth of civilization, and the rise of modernity. One problem is that we see a naturalistic phenomenon in the 'evolution of religions' and in general a dynamic that has nothing to do with religion at all.

We can discover the significance of both the Axial Age and of modernity by asking a question, Is there a second Axial Age? The rise of the modern world is simply another 'axial' transformation, disguised behind its secularism. The formulation of Karl Jaspers remains ambiguous on the question of the rise of the modern. The reason is the stumbling block created by misleading definitions of 'secularism'. Darwinism, atheism, scientific positivism, Nietzschean anti-modernism, the calamities of the First World War and the Holocaust, are all taken in evidence to either define the secular

Introduction

A New Model of History.

Our model is designed to create an outline or chronicle with a 'dynamic' built in. The result is not a theory but way to observe dynamism in action on way to some future supercomplex theory. Thus the model in essence is the same as narrative history with free agents. The basic ideas are:

1. a formalism of evolution as macro, micro, the term evolution meaning 'development' a deduction of the model by looking at the connection of evolution and history.

2. a clear distinction of the action of a system and the free agency inside it, and this will generate some meaning to the phrase 'evolution of freedom'.

3. a frequency hypothesis based on a set of intervals or epochs and their generation via a set of transitions: the resulting sequence of transitions show clear evidence of an exotic brand of directionality, hinting at teleology.

4. these transitions make sense in terms of 'relative beginnings' and suggest how an evolutionary driver can shape a larger system: their action is a set of innovations called 'eonic emergents'.

5. These transitions can be both sequential and in parallel, as relative beginnings.

6. this approach allows us to change the 'unit of analysis' from civilizations to the 'transitions' themselves.

7. all these ideas can be summarized by a contrast of 'stream' and 'sequence', as we discover an exotic form of dynamism.

8. this system evolves 'civilization' via a small subset and from different streams or civilizations a larger sequence is derived.

We will see that the only way a system this complex can work is if it is basically some kind of global or '*Gaian*' complex.

The strange properties of this model suggest a kind of minimax effect: how develop a complex variety of civilizations? A straight directionality will lose the lateral cultures. A solution is an intermittent sequence that samples different cultures in short intervals and generate a sequence from transitions in separate civilizations.

Overall, we suspect that a Great Transition stands behind the smaller sequence of transitions, and it seems that human evolution, after its stage of emergent organismic homo sapiens, moves to a larger context of cultural evolution and globalization.

or castigate it. This misses the point entirely. The 'secular' is suddenly obvious as the type of society emerging from the early modern, ca. 1500 to 1800. This is a complex dialectical spectrum (as was the Axial Age), not an 'ism' defined by some watered down version of scientism or the Enlightenment. Thus the 'secular' for us is not a philosophy, but a temporal interval in a larger sequence, with a geographical sourcing area, showing a complex dialectical center of gravity around religious transformation (the Reformation), the Scientific Revolution, emergent economic modernism (capitalism, and its potential counterpoints, e.g. socialism), the Enlightenment (and its potential/actual counterpoints, e.g. the Romantic movement), re-emergent democratic experiments, and much else. A kind of postmodern fog has already settled over our perceptions on this point.

Fig. 7 Stonehenge

1.3 History and Evolution

We are moving in two directions, and this is at first confusing. We are going to examine the realm of antiquity and the emergence of civilizations. At the same time we are introducing an at first counterintuitive discussion of evolution. As we proceed, we carry a question, what is the connection?

We are ready, to take a look at world history. Archaeological research has greatly expanded our knowledge of world history, and the result is the unexpected discovery of a mysterious dynamic generating a non-random pattern visible as the Axial Age. In fact, the scale of this process is such that we can only call it 'evolution'.

> The perception of 'evolution' in world history springs from the way we see the 'evolution' of such factors as religion in action and this is entirely different from the speculative versions of evolutionary psychology.

Thus, for the first time we can detect the unmistakable evidence of non-random evolution, and this in world history itself. This leaves us with the question, What is evolution? And this forces another, long overdue, What is the relationship between history and evolution? This could be recast as the paradoxical question, When did evolution stop and history begin?

History and Evolution A paradox confronts the distinction of evolution

Introduction

Stream and Sequence

Stream and Sequence Consider the dynamics of the Greek or Israelite Axial intervals (or any other for that matter). A stream history leads up to the Axial interval and shows transformation. This transformation generates a higher level step in a greater eonic sequence. This is the 'stream and sequence' effect. We now have two levels to our account, the evolution of the stream of cultures, and the evolution of the high level sequence. And this allows us to give expression to ideas of evolutionary directionality and progress at the higher level. Or perhaps progression would be a better word. However, the idea of an eonic sequence allows us to proceed without committing ourselves on generalizations about progress which always end up confronted with various contradictions.

Transition And Oikoumene We need one more idea to describe our eonic series, as we look at the complement to our transitions, the oikoumenes they create. And this leads us naturally into the question of the 'mideonic periods', where the center of gravity of history lies. We will attempt below to rewrite our eonic system as an 'evolution of freedom'. But note that in a system 'evolving freedom' the system must finally switch off to allow freedom to develop outside the field of system action. We can see that the initial results in the mideonic periods are mixed at best.

How Evolve Civilization(s)? The eonic sequence shows an ingenious way to 'evolve' civilization(s). The whole is too large, work on a series of localized regions. These in turn generate a set of oikoumenes or diffusion fields. The Axial religions begin to spawn universal trans-cultural diffusion fields, armed with literatures able to apply across cultural boundaries, although as we can see, the Old Testament is a curiously sluggish mixture of particularized culture elements pressed into service for ecumenical purposes.

The Frontier Effect A key property of our eonic pattern is the 'acorn or frontier effect'. The sequence restarts in a new place each time, like an acorn, just at the frontier of its predecessor. The world of Canaan, spawning 'Israel', does not look like a frontier now, but in the era of the mythical Abraham it certainly was, and we even have a 'pioneer' story about his leaving the city of Ur in a prime diffusion source, the world of prior Sumer. Greece and Rome in the Axial period were definitely still frontier areas, relative to the by then ancient world of Egypt and Mesopotamia. Each of our transitions creates a hotspot, then expands to create a new civilization, better, oikoumene. Cultural acorns sprout in this field, and then at the next cycle one of them becomes a new transition.

and history: when did evolution stop and history begin? This odd question is the clue to seeing that history and evolution must show an interconnection. Further this braiding together is likely to show a series of transitions between the two. With this clue we can rapidly find the evidence for just this, which we call the 'eonic effect'.

A moment's reflection will tell us that no instantaneous passage between the two is plausible and that our terms have been left ragged. We must, by this logic, be able to detect a Transition between evolution and history. Can we find evidence to match this deduction? Indeed, we can, our non-random pattern, the eonic effect. In fact we can say more: if we apply that same logic to our Transition we should expect it to take the form of a series of transitions in an alternation between evolution and history, as if overlayed, the one emerging from the other. The eonic effect shows just this property of transitions in a series. Have we reached the end of this Great Transition? If not, then our evolution still constitutes our present and future. We should ask who man is, with such wisdom as would constitute achievement of the title, *homo sapiens*.

Fig. 8 'The Pyramids of Sakkarah

We are so accustomed to Darwinian or reductionist definitions of genetic evolution that we forget the meaning of the term: evidence of developmental emergence by any process or dynamic. By that definition history shows a clear pattern of non-random evolution in the development of civilization (and the parallel development of human individuality).

As we proceed in search of history we will discover an irony, which is that we will find evolution in history, and then history in evolution, and this will give us an insight into the descent of man. We must move beyond the myth of purely genetic evolution, and the fixation on natural selection. We can recalibrate our definition of 'evolution' to include man's past, present, and future, with a new kind of model that can carefully define the nature of our evolving freedom.

The study of the Axial Age becomes, ironically, a study of evolution because the term applies to any perception of development. But then what is the relationship to evolution in deep time? We will suggest an isomorphism of contexts that can share a basic 'evolution' formalism, macro/micro.

Introduction

Visions of a Ghostseer

The labyrinth of modern thought is a difficult one in which the unforgiving complexities of parallel dialectical movement, seen in the divergence of idealism and materialism, can leave understanding stranded in the restricted movement of divorced specializations, and paradigms. Issues of 'materialism' and 'idealism' can vitiate thought, and deserve to be relegated temporarily to the sidelines, so that a practical study can get underway. It is important to consider the often neglected potential of so-called 'transcendental idealism', in its Kantian version. Neither transcendental, nor quite an idealism, it is the perfect complement to Newton. This crude but effective kludge is, at the least, the perfect way to state our problem, whatever its solution.

Whatever the case, the stance of science is appropriate, and a rough and ready 'materialistic phenomenology' can be our starting point

it is strangely forgotten that Kant, issues of his idealism apart, with Newton at his fingertips, pronounced skeptical judgment over assumptions, material or otherwise, arbitrarily made about the 'Big Three', divinity, soul, and free will His early essay, *Visions of a Ghostseer*, with its critique of mysticism, prefigured this classic treatment of metaphysics later addressed in his famous *Critique of Pure Reason*. The *Preface* to that critique opens with the famous statement,

> Human reason has the peculiar fate in one species of its cognitions that it is burdened with questions that it cannot dismiss, since they are given to it as problems by the nature of reason itself, but which it also cannot answer, since they transcend every capacity of human reason.

The Darwin debate can be taken as fully in the grip of this peculiar fate. This passage has suffered a strange fate itself. It was a challenge to metaphysics. Yet now science denounces Kant as metaphysical even as it makes the mistake indicated in Kant's Preface. Reductionist evolution based on natural selection is as metaphysical as it gets. If Kant is seen to be wrong somewhere, we default back to this paragraph, with no science of metaphysics, and hence no science of evolution, physics generally managing to fend for itself.

We base our discussion on the famous challenge of Kant, and the books in question make the claim to have resolved Kant's Challenge, and deserve some discussion on this basis. Academic culture is dysfunctional in the context of the domination by Darwin's theory of natural selection. Before we can do anything we need to 'debrief' darwinism as a theory.

We can start with the issue of randomness by citing 'Fisher's Lament', a perception of historical randomness, and then what we call 'Kant's Challenge' from his classic essay introducing the philosophy of history. We will move in two or more directions and the discover the larger connection between them. The model presented in WHEE deals with the larger question of world history as a whole, with the Axial Age 'inside' it. The data of the Axial Age is like one of the clues that allowed the decipherment of hieroglyphics, or a 'crib' that can give us a hint in solving a code. One tantalizing fragment allows us to consider what we are missing with Darwinism. The appearance of a non-random pattern dead-center in world history gives a hint toward solving a problem that has defeated biologists and led to the confusing paradigm of Darwinists. The issue of scale bedevils the study of evolution. In world history we are dealing with intervals of ten millennia, and far less than that if we insist on data at the level of centuries or less. As we look backward into deep time we assume we can assess intervals of tens of millennia. But that is not the case.

Fig. 9 Sumerian text: 'Gifts of Adab'

The idea of the Axial Age situated at a point which is unique in fashioning humanity both expresses the spectacular effects of the period, but, more controversially, its uniqueness. We will consider the idea of 'relative transformations' to caution the idea of this uniqueness.

We can certainly infer evolution as a fact, but we can't easily assess the dynamics, and the suspicion arises that the key actions are too fast acting to be detectable. The ten thousand years of world history, in the context of the scale of deep time, would be completely invisible. The Axial Age is a reminder that massive changes can occur under the radar of scales matching greater evolution since the dawn of life. This argument is not conclusive, since the scales of generation in deep time would not correspond to those much later, especially since we are discussing cultural evolution, not just the evolution of organisms. The scales of the Cambrian, of the emergence of hominids and

Introduction

One of the strangest aspects of the emergence of Darwinism is the sudden appearance of Alfred Wallace on the scene, triggering the publication of Darwin's *Origin*. A closer look leaves us with the suspicion that Wallace's letters suddenly cured Darwin of his 'evolution' writer's block, and ignited the cribbed notes of his Origin. The long delay in Darwin's work here has always been something of a mystery, as if he remained unsure of the basis of his claims. This story of the rigged priority upon receipt of the famous Ternate letter leaves an ambiguity at the threshold of Darwinism. Any evaluation of Darwin and his theory should consider the motives of personal ambition at the onset. And any testimony to evolution should consider Wallace's 'second opinion' on the subject of evolution, for he quite intelligently saw the problems arising with the question of human evolution.

Wallace is notorious for his later interest in spiritualism. The attempts to proceed scientifically in this area seem ludicrous to us now, and yet the question will not die in so far as Darwinian thinking cannot produce a viable definition of the organism, certainly not of man. Is the organismic totality a purely space-time entity? Even such a simple question eludes easy answer. It founders at the limits of metaphysics.

> **Just So (Ghost) Stories** It is ironic that the onset of one of the greatest critiques of metaphysics began with Kant's *Visions Of a Ghostseer*, sounding the caution that questions divinity, soul, and free will would prove intractable to scientific analysis. Darwinism gets itself in trouble on all three of these classic issues. We might smile at Wallace the table-rapper, but sound science can provide no proof against the reality of ghosts, a dismal circumstance. At least we can be sure that if such exist, Darwinism is falsified on the spot, the difficulty of ghostly forms adapting to their environment by natural selection being evident.

of history since the Paleolithic are not the same. Nonetheless, the evidence of world history is a caution. And the hint we are given warns us that random evolution, so dogmatically taken by current biology, is misleading us.

Biologists themselves have a related solution to this kind of problem with their root-model of 'punctuated equilibrium' but somehow the warning that punctuations might be real but short acting has been excised from that provocative term. We should be careful here: the term is very tempting, but its actual context is the evocative stringing together of two words from a dictionary. The term itself is a semantic orphan as if a restatement of Newton's second law (all hobbits everywhere are accelerated by punctuations to their bodies until then following Newton's first law as couch potatoes...) or more generally the principle of sufficient reason. The real value of the concept is as a neologism to bring in discussions of discontinuity, the source of outright paranoia in Darwinists. It is nonetheless true that the 'macro effect' is probably the best example there is of a 'punctuated equilibrium'.

Fig. 10 J. B. Lamarck

The term 'Axial Age' is a subject to charges of being a myth. And we have in the original WHEE already used the term on the way to a change of terminology. But the concept, with a slight revision, is quite adequate on its own, but its significance is better understood in a larger context. Jaspers defines the Axial Age as from -800 BCE to -200 BCE, but that is too long.

> The Axial Age is not really an 'age' but the transition to 'new age'. We will refer to the interval from -900 BCE to -600 BCE as the 'revised' Axial Age or interval, like a countdown, the period from -600 to -400 as the 'lift off' phase, with an analogy to rocket science. The history recorded in the Old Testament, for example, fits this new framework perfectly, with the period ca. -600 showing the onset of an almost complete new religion.

Here a close reading of Jaspers shows the way he is speaking of 'axis' points, possibly in the plural. He clearly goes in search of other axis points, and correctly but inconclusively finds the right zones, earlier Sumerian range civilizations, and the rise of modernity. But these large regions/time spans can't be axis points. The solution lies in the concept of a transition.

> Jaspers plunges us into the issues of Christian theology and ties the Axial Age to Christian history and then extends the range to a universal

Defining 'Evolution'

The use of the term 'evolution' in world history will be a stumbling block for some, even as they accept its 'Social Darwinist' usage in that context. We will settle the question, 'by definition': the word comes from *'evolvere'*, 'rolling out', and is appropriate for evidence of developmental sequences, whether in deep time, or in history. This definition is not inherently genetic, and the study of history will make clear that 'evolution' operates at a higher level than the genomic. As we move to examine world history, we discover that the non-random patterns it exhibits, as with the Axial Age, are best described as 'evolution', by definition. This usage then provokes a suspicion that what we find in history is also the case for the earlier 'history', i.e. evolutionary emergence, of man as *homo sapiens*. And the evidence for a 'great explosion' at the dawn of human speciation is tantalizing. We need not jump to any conclusion, but we must demand that Darwinian assumptions be withdrawn: they are speculative, and less plausible.

Our usage will be 'Janus-faced', with 'history emerging from evolution' (like a student graduating from school) as the 'evolution of freedom' creates a free agent who steps beyond evolutionary passivity into historical free agency. Thus 'evolution' and 'history' overlap. Consider the visual metaphor in the endnotes.   DMR

This is disorienting, but completely logical. 'Evolution' means 'evidence of development'. Our case can be qualified as 'civilizational development or evolution', or civ-ev. The question then arises, if civ-ev corresponds to the 'evolution in deep time' , DT-ev, in general or human evolution in particular, hum-ev.

This usage is empirical, not theoretical, but corresponds to a very simple model of 'discrete transitions'.

humanity. He suddenly extends the idea of 'axis' to include the period of Jesus. We will try to show how emergent religions can be tied to the Axial Interval yet generate in the wake of that source interval.

We must be clear at the start that we are expanding our account to a modernistic cosmic level and the issue of divinity goes into semantic recalibration. The Israelites thought anything that could change the course of history required the concept of 'god' or more directly IHVH. Even as we proceed via a path of Biblical Criticism in an era of science, we are confronted with the scientific equivalent of a 'something that can change the course of history'. But the divinity in question won't apply! The 'religion' in question is the output of a system that is mysterious but which resembles 'evolution' in a system's model. One that skirts uncomfortably close to a design argument.

Fig. 11 Darwin

The question of design has bedeviled theories of evolution, and our study of world history is no exception. With a new twist: a new religion with a designer emerges claiming with self-referential irony to be both the input and the output. Our model will attempt to remain within 'secular' boundaries, we allow divinities only as output. But there's a catch: our system is so close to the logical equivalent to a 'designer' that we are left with a super-puzzle. In any case we should note that the ancient Israelites saw fit to eschew 'god talk', leaving the reference in a glyph unspoken in an elegant silence.

One problem is that we can't easily rebrand the concepts behind the 'Axial Age' in the confused literature using the term. We have revised the definition nonetheless on the way to a new concept. There is an immense literature almost all of its confused to the point of incoherence. That was the reason WHEE generalized the discussion to a new framework using new terms like 'the eonic or macro' effect with a set of numerical coordinates in a series. The first edition spoke only of 'transitions' with coordinates 'ET1,2...'. That was not exactly intuitive. But the term 'Axial Age' refers to a too fuzzy entity. Here we will linger with Jaspers' term to create a heuristic discussion. That means that some readers might try to interject design arguments, including 'god' agents for dynamical discussions. Jaspers' perspective is complex but clearly invokes Christian theology. We need to find a way to deal with that. We will therefore continue with our sideline discussion of the work of J. G. Bennett's framework in *The Dramatic Universe*: his work is useful in the

Introduction

Problems with random evolution, and How would we detect teleology?

The perspective of Darwinism is that of random evolution, and this framework has always concealed a host of problems, however attractive the concept is for proponents of reductionist science. Random evolution

1. must skirt severe improbability, as the scientist Fred Hoyle warned,

2. overcome without a template, system memory, or feedback control the inherent tendency to peter out, deviate, or retrogress,

3. operate in partial steps to construct complex objects at random, with no direct connections between steps, in constructs with tens of thousands of parts,

4. effect infinitesimal, geographically isolated innovations into species level change over large regions or whole species.

This is but a short list. It should remind us that Darwinism is implausible from the start, and yet seems to be scientific because the fantasy of natural selection is never tested against reality and thus avoids the really difficult implications in our list of problems. At the same time, our four problems point to something that must be complex beyond our understanding. It is not surprising biologists cling to an oversimplification like natural selection that makes these difficulties vanish.

Detecting Teleology It is not hard to deduce what evolution should look like from these difficulties, which must leave their signature in the data of any given chronicle. The problem is that these issues imply something controversial: teleological sequences. What form would teleology take, and how would we recognize it?

There are very few solutions to this set of contradictions: one is that of an explicit evolutionary driver, a sort of macro process that operates *intermittently* over the long range, and acts on wholes via transitional areas of reasonable size. That's a tall order. But sure enough world history will give us an example.

way it produces design agents in nature as 'demiurgic powers'. Whatever the status of this, it can help to distinguish 'spiritual' beings in nature, rather than a transcendent entity beyond nature. This distinction, with a dash of Bennett's near Sci-Fi account is invaluable to clarify terms.

We must be careful to counsel this issue of theism: the critique of religious historicism, as we explore history, will be to set the record straight here, and there is a curious irony: the evidence of a higher power in history becomes overwhelming, but that isn't the result of 'god' doing something. Debates over atheism/theism and the Biblical history are confounded by the new model. Jaspers sensed this: a global riddle replaces the localized Biblical tale. The term 'Axial Age' is a maiden in distress and enters self-falsification at the first attempt to explain it. Two sides, religious and 'secular' suffer collapse of their views. This should be no problem. Have we forgotten the reluctance of religious greater antiquity to suffer 'names of god' and that this entered the tradition of the Israelites. The religion of IHVH became the religion of Jehovah, and the latter has suffered the criticism of secularists while the former bespeaks the silence of the mysterious in the emergence of, well, 'axis' points.

Fig. 12 Wallace

> The confusion over the Axial Age rises from the lack of concepts to deal with large-scale 'motions' of civilizations: we need a simple model to stabilize this confusion:

> We will bypass the concept of civilization with an idea of 'transitions', as time-slices or relative transformations. We will distinguish thus the 'stream' of a civilization and the 'transitions' which create a meta-sequence. These transitions are finite intervals with a 'divide' or closing point. The action of individuals inside larger dynamical systems forces a distinction of free action, and system action, and this applies in all sorts of ways, viz. a sheperd's action on a herd, and the 'free action' of the herd itself. Instead of civilizations we will think in terms of the 'transitions' and the new oikoumenes they generate. This core of ideas can help to resolve the 'blur' of historical visualization that leaves the historical drama so seeming incoherent. We will develop the model as we go along, but a point of reference is the generalized model of 'punctuated equilibrium'.

The book is based on the fact that the *Introduction* to WHEE is really almost a book in itself with a focus on the Axial Age and this can be the core

Introduction

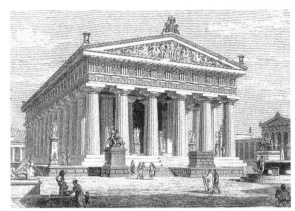

Fig. 13 Ancient Greek temple of Poseidon at Paestum

Non-genetic Evolution

The Axial Age is a clear and devastating challenge to ideas of natural selection and of genetic evolution. Darwin is more or less on record as assuming that natural selection is at work in the destruction of primitive races and that the achievements of the Greek classical period are the result of differential natural selection, a most doubtful viewpoint. Why was there a Greek flowering of culture? Because, by natural selection, the Greeks were smarter or some superior race? What about the Hittites? These were essentially the same tribal and linguistic stock. Yet they shew very little creative culture. They weren't in the macro mainline. What about the Romans? They are almost a variant tribe, yet already look backwards to an established tradition. One is just before, the other just after. In parallel we find the post-Vedic mimic in concert the Greeks in music of different key. This has to be a problem of periodization. The foundations of the Greek classical achievement appeared at almost record speed from -900 to -600 for reasons, we can strongly suggest, that were conditioned by zone and period, in a master sequence. It is a question of macro determination. This remarkable interval, echoed in the raw structure of the Old Testament, has no other account than as a 'fast interrupt'. Even if we thought they had special talents or intelligence as a culture, this other explanation would hold good. For we will move to see the full counter-experiments in all combinations, the comparable Hittites, and (Greek) Mycenaeans before, the Romans just after. In general, evolutionary theory assumes that selection for intelligence is a foregone conclusion in the evolution of the brain. Even the small snapshot we have of human history shows the 'survivors' too often to be a very restricted range of men. Uphill selection requires unique conditions for success.

of our new collation with new material of interpretation on both the Axial Age and the evolution question. This can in turn be connected to questions of human evolution and the companion volume, *Descent of Man Revisited*.

Books on the Axial Age have consistently misinterpreted the phenomenon, and the seeds of this confusion spring from the original text of Karl Jaspers, whose remarkable study nonetheless pointed to a remarkable discovery.

Fig. 14 The Parthenon in art

It both initiated the study of the phenomenon, and fixed its interpretation into a distinctive mode somewhere between the religious philosophy of history and the borderland of a secular sociology of religion. Jaspers thus failed to see the place of Axial Age Greece in the study of the larger question, but he was one of few, in the seemingly contradictory world of a Christian theologian and a student of Nietzsche. Jaspers' work has been followed by a considerable number of studies, all of which in the author's opinion are entirely inadequate.

More generally a disembodied observer of world history with the power to influence historical streams could create a sequence from short intervals in parallel streams. The phase of Axial Greece seems to jump to the phase of modernity. Remarkable. The existence of parallel synchronous intervals complicates the discussion, but these can all contribute to the next step! Thus, to a close look, China, India, Persia, 'Israel' all contribute to the interval generating modernity:

The challenge to Darwinism in WHEE has led to the attempt by authors such as Robert Bellah to refix the Axial Age question so that it won't disturb the flatland platitudes of Darwinian orthodoxy. The two works he has produced, one a collation of scholarly essays, are notably discourteous for trying to scotch the perspective of WHEE without any explicit mention of this work. These pretentious works with a kind of academic superiority complex are actually filled with elementary blunders. The result is an indication of the way scholarly opinion attempts to enforce the basic paradigms of evolution, and the sociology of religion. The result is a distortion of many issues, among them the question of Indian religion including Buddhism. Students of the sociology of religion have struck out on the issue of the Axial period despite

Introduction

Our data generates an overwhelming sense of design. We found this by using systems analysis, not by theistic speculation. Compare the account of the Old Testament with the plain macro analysis of Archaic Greece: the latter is far superior. And our sense of design is global, and transcends of the myths of revelation of the monotheisitic Axial Age religions. That said, the Israelites detected the macro effect.

Note that this design sense applies with especial force to the modern transition, its ideologies, and revolutionary politics. The traditionalist anti-modernism of much religious design mythology is seen to be archaic confusion. We must reinvent design interpretation via a neutral systems analysis. The philosopher Hegel was the first to sense this, but produced a systematics that is arguably metaphysical. Our better approach carries his insight to a deeper level, and shows how modernity is really another 'Axial Age'.

Our model resolves the beautiful way in which the religions of the Axial Age spring as from seed to blossom in the middle period, Christianity and Islam, and in Buddhism in the Orient. The close connection can be seen in the way the phase of Mahayana and the rise of redemptive Christianity are synchronous in phase and connected in content. The enigma of the 'god/man' in Christianity has been the source of endless confusions, but is a simple and elegant symbolism of the action of the macro and micro in a meeting in time.

These religions are Axial Age productions, prodigious in scope, and are perhaps destined for recasting in the wake of the modern transition. The attempt to replace them with modern scientism is not likely to succeed.

We should also consider the existence of demiurgic powers in the realm of nature, as a side hypothesis. A strong suspicion arises of the existence of life-forms beyond the framework of body-mind climaxed in man. This remains a mystery for future understanding, and this thinking can, in any case, be a lightening rod against false speculative theism which as we can see has distorted the Old Testament's account with primitive design logic. The reality is almost more spectacular.

Jaspers' clear trek to the threshold of the key to the mystery.

The only way to deal with the data of the Axial Age is with a simple model that uses periodization to show the action of a dynamic in a longer series. The dynamic is a chain of discrete transitions, sequential and in parallel. We are in a situation where science must confront personal knowledge, to use a phrase from a classic text of that name by Michael Polanyi. A solution to a physics problem is shared by a community of scientists. But we have a data zone so vast as to ask reading ten thousand books, to start. But for one book we would be lucky to manage a thousand, and a community of historians would not follow the same course of reading: the result with this kind of model is a no longer fully shared experience.

Fig. 15 Colossi of Memnon

The stunning elegance of this type of system is belied by the need to read books on multiple separated zones of world history. That is hard, but highly instructive. A model of this type is not like a deterministic system with a closed equation or 'law'. It defaults to an outline in which the data must be described at each point with a answer to the question, 'what happened'! That's called history and it takes the form of a narrative! This model can resolve the paradox of a science of history, which is not possible in the classic sense of laws of physics. But there can be structure of many kinds instead, and these are found by the evidence of something 'non-random' in the background.

Attempts to interpret the Axial Age by students of religion have also mostly failed to grasp the perspective of Jaspers, who in turn could not quite see the broader implications of his own study. But Jaspers' classic stages a contradiction that won't go away until a new perspective is found. The overall situation is of secular thinking exposing theological historicism and a new brand of the philosophy of history, with newly discovered theological implications, subjecting Weber's 'Iron Cage' flatlanders to a kind of shock treatment.

The questions of religion and secularism in modern times, and here Jaspers' correctly resisted the confusions of Nietzsche, have stumbled into an impasse. The riddle of modernity has confounded both religionists who expect

Introduction

The Axial Age: Religion, macro and micro

The phenomenon of the Axial Age shows us the solution to the riddle of evolution, but instead has produced a whole series of false interpretations. The only way out of the morass is to consider our frequency hypothesis taking the data as a set of discontinuities in a timed sequence. Then we must carefully study the differentiation of effects in different cultures. It is not a 'common philosophy' applied in different ways, but parallel transforms of source areas. To try and find a common denominator as an 'Axial Age' philosophy won't work. We see contrasting opposites and a balance of diversities, increasing the future potential of the system. The 'evolution' of religion is powerfully illustrated in the way a 'macro' effect takes up the streams of religious culture and amplifies them, in two cases, India and Israel/Persia, into what will become materials for world religions. The Indian case is especially significant because a tradition of great antiquity, the so-called Jain, remorphs on schedule into Buddhism, in the wake of the terminating sequence of Teertankers, concluding with Mahavir! The sudden coalescence of Persian and Israelite monotheisms at the conclusion of the Axial interval (by our measure) is a spectacular effect, leaving sociological causation theory far behind. As the Israelites well knew there was a higher dimension to what befell them.

It nonetheless remains the case that Archaic Greece, our putative source of modern secularism (but in a flowering of polytheism as 'art religion'), is the clearest exemplar of the Axial effect. Its massive cluster of innovations coming in and going out with a spookily exact schedule is far more 'miraculous' that anything portrayed in the primitive Old Testament. The riddle of Christianity and Islam show a beautiful resolution as Axial Age seeds come to full bloom in the 'middle period' of our sequential series.

with postmodern confidence this era will pass and allow the restoration of antiquity, and the secularists themselves who cannot sort out the complexity of the modern transformation with the tools. Mostly those of economic sociology, given them by the scientific perspective. We have already suggested that the Darwinian axioms have led them into the confusions of scientism, with the assumptions that theoretical causality must resolved the problem of history. We attempt a different perspective: apply a framework of universal history, as attempted first by the philosopher Kant, to construct the bridge between religious and putative scientific histories.

Fig. 16 Ancient Ruins -Philae

Notes

1.3.1 The Legacy of Darwinism

At a time when theories of evolution are under renewed controversy, discussion is hampered by the remoteness of the phenomenon of evolution, and the use of indirect inference to speculate about deep time. In the face of much criticism from religious Creationists, now accompanied by the Intelligent Design movement, adherents of Darwinism forever defend a flawed theory that has been challenged from its first appearance. The objections of the first reviewers of Darwin's book, indeed even of T. H. Huxley, the original champion of the theory, were never quite answered in the tide of paradigm change that swept modern culture. The perennial issue is natural selection as the mechanism of evolution. The assumption that evolution occurs, and must occur, at random is the crux of the dispute, one unreasonably confused by the claims of religion versus science.[1]

---

1 Ernst Mayr, *One Long Argument: Charles Darwin and the Genesis of Modern Evolutionary Thought* (Cambridge: Harvard University Press, 1991), F. Hoyle & N. Wickramasinghe, *Evolution From Space* (London: Dent, 1981), Robert Reid, *Evolutionary Theory, The*

Introduction

Random Evolution: Climbing Mount Improbable?

One of the most confused claims made by Darwinists concerns the randomness of evolution by natural selection. It is obvious that Darwin's theory is about evolution by accident, but since the improbability of this begins to demand some account we are given a revision in the works of Richard Dawkins where it is said that while mutation is random, natural selection is non-random. This odd way of restating Darwinian assumptions about chance is a suspiciously convenient change in the original meaning of the terms used, and seems little more than a rhetorical finesse designed to throw critics off guard. As Dawkins notes in *Climbing Mount Improbable*, "It is grindingly, creakingly, crashingly obvious that if Darwinism were really a theory of chance, it couldn't work. You don't need to be a mathematician or physicist to that an eye or a haemoglobin molecule would take from here to infinity to self-assemble by sheer higgledy-piggledy luck." But it is quite as obvious that Darwin's theory is one of chance, so we are done. Richard Dawkins, *Climbing Mount Improbable* (New York: Norton, 1996).

> World history gives us a stunning example of non-random evolution in a series of beats or waves stretching over many millennia.

Dawkins proposes that the problem is resolved by the accumulation of small steps, then bets his argument on a completely incorrect analogy to computer programming. Again, as Hoyle observes, chance wouldn't even get a single polypeptide straight, and nothing in genetic programming has ever solved this problem. Beyond the hype, it would cause a feeding frenzy in the stock market if any computer program was found to do what is claimed. It would revolutionize industry. We would certainly know that this was the case! Instead we see a sheepishly heuristic wish fulfilment at work in the Darwinian mythological fantasy world.

The rise of molecular biology shows a complexity of structure that cannot easily survive statistical challenges to claims of random emergence. The new genetics and the emergence of developmental biology have exposed the limits of Darwin's original theory, in the remarkable findings of complex

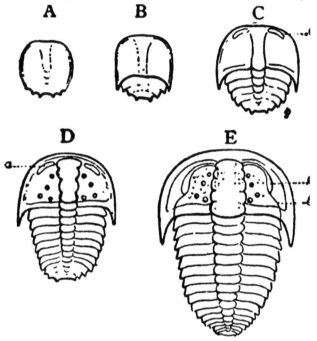

Fig. 17 Trilobite evolution

biochemical systems and evo-devo. Therefore the critics, whatever the public pronouncements of Darwinists, have essentially won the debate, and retabled the views of many of Darwin's predecessors at the birth of embryology in the generation before *Origin*. We might proceed on that basis, beyond the

---

Unfinished Synthesis (New York: Cornell, 1985), Robert Wesson, *Beyond Natural Selection* (Cambridge: MIT, 1991), Michael Denton, *Evolution: A Theory in Crisis* (New York: Adler & Adler, 1985), William Dembski, *No Free Lunch* (New York: Rowman & Littlefield, 2002), Lee Spetner, *Not By Chance* (New York: Judaica Press, 1998), Robert Behe, *Darwin's Black Box* (New York: Free Press, 1996). Stuart Kauffman, *At Home in the Universe* (New York: Oxford University Press, 1995), Johnjoe McFadden, *Quantum Evolution* (New York: Norton, 2002).

A useful critical history of Darwinism can be found in Soren Lovtrup, *Darwinism: Refutation of a Myth* (New York: Croom Helm, 1987). Lovtrup notes, "I believe that one day the Darwinian myth will be ranked the greatest deceit in the history of science. When this happens many people will pose the question: How did this ever happen?" Soren Lovtrup, *Darwinism: Refutation of a Myth*, p. 422.

Introduction

Beyond Natural Selection

The most confusing aspect of the study of evolution is the nature of the first step, natural selection. The debate over evolution tends to degenerate into a conflict of science and religion, deflecting our attention from the basic problem with Darwin's theory: the limits of selectionist explanation with 'Just So Stories', or adaptationist scenarios. It is very convenient for Darwinists to confront Creationist critics who tend to reject the fact of evolution. This deflects attention from the real problem. In the final analysis the proposition of natural selection would seem implausible. The original criticisms of the first generation of Darwin critics in many ways still stand. T. H. Huxley himself, ironically, warned Darwin on the eve of publication of the problem with natural selection. The intractable character of the debate is no mystery and arises from the violation of the limits of observation, Karl Popper famous 'metaphysical research program'.In general, severe, almost certainly fatal, mathematical challenges have always stood in the way of selectionist assumptions. In a now classic text, *Evolution From Space*, Hoyle and Wickramasinghe give one version of this objection.

> Darwinian evolution is most unlikely to get even one polypeptide right, let alone the thousands on which living cells depend for their survival. This situation is well known to geneticists and yet nobody seems prepared to blow the whistle on the theory. Cf. F. Hoyle & N. Wickrmasinghe, *Evolution From Space* (London: Dent, 1981), p. 148.

distracting cultural politics of evolutionary theories, which now sees the resurfacing of the design theology of the generation of Paley. Nothing in the methodology of science requires us to accept the claims of natural selection as established.

The Developmental Perspective Although the findings of so-called 'evo-devo' have already been grafted onto the mythology of natural selection, they raise the question of developmental interpretations of evolution, thence of natural teleology. As we examine world history in light of the eonic effect a developmental sequence unconnected with genetics emerges with a demonstration of evolutionary directionality visible as macroevolution over five millennia. The representation of teleology as intermittent directionality suddenly gives meaning to the idea of 'punctuated equilibrium'. World history has its own 'evo-devo', with no connection to genetics.

Fig. 18 Early Pleistocene animals

The new developmental perspective, although essentially genetic, strengthens once again our suspicion of processes that go beyond the selectionist account. The problem is one of observation. Evolution at close range is very difficult to observe. Darwinism applies a universal generalization to unseen events and claims in advance of demonstration that natural selection is the mechanism, frequently on the basis of no observations at all. As if Newton's second law were taken forth from physics, Darwinism assumes no differential transformations at short intervals are to be found in the immense interstices of time they take for granted. Was this a theory or the absence of one? [2]

---

[2] Sean Carroll et al., *From DNA to Diversity* (New York: Blackwell, 2001), Rudolf Raff, *The Shape of Life* (Chicago: University of Chicago, 1996), J. Gerhart & M. Kirschner, *Cells, Embryos, and Evolution* (New York: Blackwell, 1997), Jeffrey Schwarz, *Sudden Origins* (New York: Wiley, 1999), G. Miller & S. Newman, *Origination of Organismic Form* (Cambridge: MIT Press, 2002).

Introduction

The Limits of Observation: Observing Speciation?

Darwinian speculation greatly underestimates the difficulty of observing evolution, and tends to substitute assumptions about natural selection for the hard work of observing evolution in action. Once we really begin to observe 'evolution' we see that it is a non-random process that stands out against the backdrop of deep time. The Hurricane Argument shows the problem with 'jungle surface' observations of life (the source for Darwin/Wallace of their theories). That surface suggests natural selection. But the reality of speciation is 'seen' only over millions of years in diverse sections of a global environment. Not surprising the problem is confusing. Darwin's theory is a wild guess applied to the immense vistas of deep time. Those unobserved intervals can fool us badly. One way to see the problem with claims for natural selection (which is, of course, always present) is to look at history, another to consider the way meteorologists study weather.

The Hurricane Argument

Consider a hurricane, a very brief event by comparison, as a global 'system evolution' on the surface of a planet. We know a hurricane when we see one, but its dynamics, mechanism, and full progression require incremental 'closing' on degrees of evidence and observation, a task not fully accomplished until the advent of satellites able to map global coordinates. In the same way we know evolution when we see it, roughly speaking, given the fossil evidence, but its dynamics, mechanism and full progression require incremental 'closing' on degrees of evidence and observation, a task not fully accomplished. Note the analogy suggests global positioning satellites over the entire planet over millions of years, to observe drifting species and their changes. Suppose an observer in outer space only had loosely sampled data on pre-Neolithic man, and post-twentieth century man, and then conjectured that some mutation caused this dramatic change.

This analogy shows at once where Darwinism departs from scientific practice. Historians routinely assume they must close on the facts in such an analysis, yet Darwinists wish to claim exemption. We have no fully observed datasets in Darwinian deep time. It is an insidious trap.

Evolution and Ethics The failure of causal science to explain the evolution of ethics is a striking revelation of the limits of reductionism. This is in fact an old issue, and the secular philosophical verdict of an earlier period is that science is intrinsically limited here. We should note that the philosopher Kant, already from the generation after Newton, was about the business of correcting this reductionist confusion, witness the clear distinction in Kant of theoretical and practical reason as a way to mediate causal phenomena and intentional action. He is considered purely a philosophic outsider to science, but that is misleading. His deliberations on freedom and causality strike to the essence of what is creating the confusion over evolution.

The basic issue is that no one is under a truly scientific obligation, to take Darwin's theory of natural selection as established, or grounds for the blanket revision of all views of man and culture. Back to square one: an operational hypothesis. Most importantly, this is not the same as denying the 'fact' of evolution. But what are the facts pertaining to the descent of man? We have a very weak empirical record here. Darwin's oversimplification succeeded as a bestseller, but a host of critics realized almost at once a problem with the basic claims. And we now have the Darwin book market where the calculation of dissent on sales causes amusingly undisguised Darwin prostration. This drives out clear exposition of the facts. New findings are disguised behind Darwin eulogies. Contradictory issues are finessed in double talk.

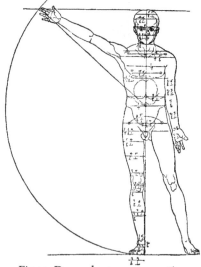
Fig 19. Durer: human proportions

> The Darwin debate has assumed a new form in the so-called Intelligent Design movement, which has resurrected the world of Paley, and the obsessive dialectic of theists and atheists heats up once again. The argument by design has a long history, and this is not the same as the issue of 'design' as such. It is not hard to see that 'something like design' is at work in genetic structures. Design arguments tend to confuse two meanings of the term 'design'. It is incontestable that many biochemical structures show design, in the complexity of their almost programmatic functionality. We might call G-design the action of a known 'designer', viz. a supernatural agent (god?), with the term N-design to refer to

Introduction

Punctuated Equilibrium and 'Gaps' Arguments...

The Darwin controversy frequently breaks down into a debate over continuous or discontinuous evolution. The foundation for all claims about evolution lies in the fossil record. But the question of the fossil record is not so simple. One of the most persistent criticisms of Darwin has always been that of the so-called 'gaps' in this record. Over and over we see the phenomenon of rapid emergence followed by relative stasis.

Here critics of Darwin have too often fallen into confusion themselves, because the whole idea of a 'gap' in the record suffers from mis-definition, if not incoherence. Fatal theological temptations induce hallucination here in many otherwise sincere minds aware of the problems of the fossil accounts. Although it is certainly true that the fossil record is very sparse, too sparse to maintain Darwinian certainties, it is not likely that one will find 'gaps' in the record. What is a gap? It is highly likely that there is a continuous sequence of organisms showing an unbroken lineage of bodily forms. That is not the same as saying that natural selection alone is at work. But these critics have a point, and a refinement of the 'gaps' argument is easy to provide, hence the challenge to Darwin's theory remains in some form. Taken over all, without claiming gaps in the record, we should suspect that something is speeding up the process of evolution beyond the rate entailed by natural selection.

Indeed, conventional Darwinians such as S. J. Gould upgraded this argument with the various claims for so-called 'punctuated equilibrium', which amounts to seeing that emergence is often very sudden, followed by a period of stasis where the rate of change is small, or nonexistent. Granting that such data is hard to interpret, the basic issue simply won't go away. These theories suffered from the inability to disassociate themselves from the fallacies of natural selection, as they attempted to have their cake and eat it too, by proposing various 'levels of selection'. But real evolution is altogether likely to be something different. And it might well 'punctuate', this being followed by some sort of 'equilibrium'. The issue is bound up in distinctions of microevolution and so-called macroevolution, or speciation. The existence of microevolutionary processes is not in doubt, but the elusive factor of macroevolution remains unclear.

the bare functional aspect of complex biological structures. We can infer N-design, but this does not resolve the question of its evolution. It is hard to explicate N-design by arguments using natural selection. It does not follow that we can infer G-design. The design argument is ambiguous and is really a theological version of teleological thinking. In the pursuit of N-design the factor of teleology might arise as a challenge to reductionism, but this teleological aspect can better be seen as a discovery of methodological naturalism.

The stubborn persistence of the Darwin debate is therefore no mystery, and is not the result of Creationist conspiracy. The rise of Darwinism has produced a false view of man, we see the long-predicted limits of the modern scientific world view. It is easy, in the case of Darwinism, to see this if we explore the limits of theory, for example, in the realm of ethics or aesthetics. Beyond that lies the immense realm of 'potential man' clearly recorded in traditions such as those of the classic Buddhist sutras. Hardly a single reference to such discourse occurs, or is allowed, in scientific literature, a clear sign of institutional agenda. Adaptationist scenarios of the Darwinian type must endure a reality check here, yet the illusion induced by the all-explanatory theory is so ingrained none see the discordance as even odd. The claim by narrowly specialized scientists to a methodology that can pass judgment on all questions, sight unseen, in a hierarchy of credentialed expertise has become a strategy of social domination enforcing a world view that most are forced to disregard in private and assent to in public.

In a nutshell, there is, as yet, no methodologically sound basis for a theory of evolution. That's a surprising statement, but the point will become obvious as we look at the gray area between history and evolution. We should recall the reservations of Kant, as to the hope 'that one day there would arise a second Newton who would make intelligible the production of a single blade of grass in accordance with the laws of nature the mutual relations of which were not arranged by some intention'. Darwin's theory, at least, does not resolve such doubt.[3]

The Metaphysics of Evolution The philosophy of Kant offers a useful benchmark for the examination of evolutionary theories as these impinge on the intractable issues of metaphysics. Questions, he warns, of god, soul or self, and free will are destined to exhibit

---

3 W. S. Körner, *Kant* (London: Penguin, 1955), p. 197. Immanuel Kant, *Critique of Judgment*, trans. J. H. Bernhard (New York: Macmillan, 1951), p. 258. For the teleomechanists, see Timothy Lenoir, *The Strategy of Life* (Dordrecht: Reidel, 1982).

Introduction

Kant and Natural Teleology

As biological science in the Newtonian legacy emerges in the era of positivism the denaturing of teleological components leaves Darwinists stranded with no definition of an 'organism'. This situation was virtually prophesied by Kant whose work suggests issues of natural teleology. The data of our macro effect, proceeding empirically, gives us an actual example: a intermittent oscillator that expresses directionality, i.e. a hybrid of mechanical and teleological components, both and neither. But this phenomenon has a 'noumenal'/'phenomenal' Janus-face.

As Timothy Lenoir notes in *The Strategy of Life*, "Teleological thinking has been steadfastly resisted by modern biology. And yet, in nearly every area of research biologists are hard pressed to find language that does not impute purposiveness to living forms. The life of the individual organism—if not life itself, seems to make use of a variety of stratagems in achieving its purposes. But in an age when physical models dominate our imagination and when physics itself has become accustomed to uncertainty relations and complementarity, biologists have learned to live with a kind of schizophrenic language, employing terms like 'selfish genes' and 'survival machines' to describe the behavior of organisms as if they were somehow purposive yet all the while intending that they are highly complicated mechanisms. The present study treats a period in the history of the life sciences when the imputation of purposiveness to biological organization was not regarded as an embarrassment but rather an accepted fact, and when the principal goal was to reap the benefits of mechanistic explanations by finding a means of incorporating them within the guidelines of a teleological framework. Whereas the history of German biology in the early nineteenth century is usually dismissed as an unfortunate era dominated by arid speculation, the present study aims to reverse that judgment by showing that a consistent, workable program of research was elaborated by a well-connected group of German biologists and that it was based squarely on the unification of teleological and mechanistic models of explanation." DMR

antinomies that will haunt any universal generalization. We have the Darwin debate in a nutshell, and can see at once that Darwinian natural selection, used as the universal talisman of metaphysical reduction, presumes judgment on unobserved totalities, and is troubled on each of these questions. Questions of divinity founder in the design debate, of soul in the basic definition of self and organism, and free will in the attempts to reduce moral action to the mechanization of adaptationism. Current biology lacks so much as a basic definition of the organism.

A clue to the problem lies in the failure to produce a science of history, where the facts are visible, even as Darwinists claim a science of evolution, where the facts are not visible. And at what point do we divide history from evolution? This situation is altogether odd, and we left suspicious Darwinism is failing a photo finish test. Not a single hard result has ever been achieved for a science of history. That should make us suspicious of Darwinian claims at the onset. We indulge in far too much idle talk about evolutionary theory in the abstract. These discussions are impoverished, but brilliant sounding speculations about something we never observe. It's time to take a long, slow motion look at the one good data set that we have, world history. We will soon be cured of Darwinian fantasies. The scale of evolution is tremendous. Even the record of world history, five thousand years over the whole surface of a planet, is nothing compared to deep time. That is a reality check. We see at once the fallacy of throwing generalizations at such a complex system. It is primitive behavior.

Fig. 20 Kant's Critique of Pure Reason

Is There a Science of History? The question of a science of history generates a contradiction that the Darwinian framework never addresses. The question is at the core of a Kantian critique of metaphysics and demands a way to reconcile the so-called antinomy of freedom and causality.

Looking at history we can easily show where Darwinian theory is going wrong. The relationship of history and evolution creates a paradox, and

Introduction

placing the two in conjunction allows us to infer something about earlier evolution. The quest for a science of history is now beginning to overflow from Darwinian confusion as a reductionist tactic for the social sciences in the claims of sociobiologists, ambitious to dismiss all other forms of discourse. It seems like a welcome mistake, a foolhardy gesture we can applaud! Just at that point we do have facts, facts that can stop Darwinist thinking in its tracks, and in the process discipline the current confusions.

Huxley's Contradiction and Evolution #1 and #2

That darwinism and similar theories don't work with history is obvious from our perspective of very late civilizational 'evolution': the issue of values is intrinsic.

We have stumbled on the subtle problem with Darwinian thinking, and the possible answer: something is producing large-scale historical change, and this isn't natural selection. Further, one of the most unfortunate consequences of Darwinism lies in its unwitting generation of Social Darwinism. Often blamed on Spencer, this ideological confusion of Darwin's theory lies squarely in the theory itself, with its emphasis on natural selection.

Huxley's Evolution # 2 It is T. H. Huxley himself who spotted the flaw in the theory of natural selection in his work, Evolution and Ethics, and in the process unwittingly exposed a paradox in the theory he had so long defended. His perception was that there must be something else beside the 'law of evolution', survival of the fittest, at work, for man was condemned to oppose its effects in practice, on ethical grounds. Whence, if we accept this dualism, comes this evolution # 2? Here the data of the eonic effect shows us at once two levels of evolutionary action. The eonic effect shows us evolution #2. WHEE

Theories of the evidence The Darwin debate constantly scrambles the issues of the 'fact' of evolution and the 'theory'. There is a complication here, which is that we can distinguish a 'theory of the evidence' from a 'theory to explain that evidence', should that theory of the evidence graduate to stable data. Darwinism has yet to produce a proper theory of the evidence, that is, it has not actually observed in full 'how evolution happens'. And this itself might require a theory, e.g. that 'evolution' shows a macro pattern. This subtle difference constantly confuses all discussion. In economics, for example, a theory of evidence would be, as a theory, that economies show cyclical behavior. A second theory to explain the first, i.e. explaining cyclical behavior, is quite another task. Note that without a detailed record we would be likely to think

Human Evolution: The Great Explosion

The evolution of man is, and remains, a complete mystery, although world history can give us important clues. There is something almost mythological in the projection of Darwinian scenarios of natural selection onto the Paleolithic. Such evidence as we have is mostly that of skeletal remains, highly incomplete, of a series of hominids stretched over millions of years. Dogmatism in such a situation takes on an almost religious character in Darwinists. In the midst of this void of hard information we are to believe that all the complex functions of the human advance are to be ascribed to processes of natural selection and adaptation. Such claims, pressed into service for metaphysical conclusions, are weak in their evidentiary basis. In contradiction to this, flagrantly out in the open, is the evidence of a Great Explosion in the period up to ca. 50,000 BC, when modern man is suddenly in evidence. As if crossing a threshold *homo sapiens* suddenly begins to leave traces of all the forms of higher culture that are characteristic of man as we find him in history. The suddenness and depth of this rapid passage, if we can trust the data, call out for explanation beyond the standard and very vague claims of mysterious mutations. This is really a question of what we mean by 'macroevolution', as opposed to 'microevolution'. Is not Darwin's theory really one of microevolution? The problem is that observing anything that resembles macroevolution demands a very detailed record of evolutionary sequences, and this invokes a crisis of correct observation. There is an irony to our views of evolution. We look to deep time to find the answers to our quest to understand evolution, and yet we have very little data to conclude anything. We then apply that thinking to history, and yet here we have what is really a far more detailed record, seen at close range. We fail to suspect the fallacy here, or that history itself shows the direct evidence of evolution.

in the abstract about economic systems. This example shows the dilemma of Darwinian theory. We have no detailed record of the way evolution actually happened, and tend deal only in abstractions based on Malthusian or other misleading examples. This is clearly the trap into which Darwin and Wallace fell, because they were struck by the teeming behavior of jungle populations with its clear profusion of speciation processes. They thought the full evolution of forms was explained by its surface aspect, the competitive struggle in biogeographical regions.

Lamarck's two-factor theory We are starting to see the need for two levels of explanation in the discussion of evolution. It is significant, and forgotten, that Lamarck, his more well known theory of adapatation apart, proposed a double aspect to evolution, progress and deviation. Rightly or wrongly, the idea of evolutionary progress is rejected now, but the more basic point about two levels to evolution remains on the table. We are left wondering how the more 'scientific' Darwinism took off with a one-dimensional oversimplification. Because pure random evolution is implausible, at least to some, one tends naturally to find two levels to evolution. If we try to eliminate one level, we always end in difficulty. The problem is the extreme difficulty of observing the higher level, and the confusion over ideologies of evolutionary progress applied to one level. But it is interesting that with a one-level theory Darwinists end up bickering over levels of selection, punctuated equilibria, and are forced to confront stasis and rapid change in alternation with no means to stuff both in the same box. Don't confuse this with Lamarck's idiosyncratic and controversial views on adaptation.

Wallace on Human Evolution

Wallace is an important, but neglected, figure in the emergence of evolutionary theory, and his views, whatever our perspective, are not refuted by anything in the spurious abuse of Darwin's theory of natural selection. Let us note, then, that one of the co-discoverers of selectionist theory later dissented on the question, as far as the descent of man is concerned. Wallace (who started as a super-selectionist) saw something that becomes obvious in light of the eonic effect, that is, the appearance not of adaptive traits, but of potential that emerges through self-realization (making the term 'evolution' ambiguous). His classic observation was that

> ...in creating the human brain, evolution has wildly overshot the mark.

> An instrument has been developed in advance of the needs of its possessor...Natural selection could only have endowed the savage with a brain a little superior to that of the ape, whereas he possesses one very little inferior to that of the average member of our learned societies....

This sentiment springs to life once we see the way Wallace's dilemma reflects on history. We are confronted with questions about the meaning of evolution, if history shows yogis exploring consciousness in traditions as old as the emergence of civilization. It is entirely possible man came into being as he is in times unseen in the Paleolithic, and that what we sense as 'evolution' is another process entirely, a kind of self-realization of potential. It is still evolution in our sense.

2. THE AXIAL AGE: A RIDDLE RESOLVED

> The history of mankind can be seen, in the large, as the realization of Nature's secret plan to bring forth a perfectly constituted state as the only condition in which the capacities of mankind can be fully developed, and also bring forth that external relation among states which is perfectly adequate to this end. From Kant on history

2.1 World History: An Undiscovered Country

We live in the first generations with enough data to tackle the question of world history, not only in the extension backward into the dawn of man, but within proximate antiquity itself where the data of the Axial Age has altered our views of the dynamics of historical emergence. This data has in turn pointed to a larger pattern of dynamical action stretching across the span of civilizations emerging from the Paleolithic. This larger pattern, which we can call the eonic or 'macro' effect, is the evidence for an evolutionary framework behind the emergence of civilizations, with its complex chords of innovations

Figs. 2.1, 2.2 First World War battlefield, Ypres

or 'eonic emergents'. The use of the term 'evolution' is controversial due to the dogmatic reign of darwinism, but the term itself refers to any process of development that shows a sequential logic of forms and this fits the data we see in world history. And what we discover suggests in fact a canonical version of the 'evolutionary' with a further suggestion which won't go away that this is directly related to the earlier evolution of man at the dawn of *homo sapiens*. Our basic strategy is to observe the remarkable non-random

pattern in world history and to create a simple type of model to explicate its

A Riddle Resolved

The Introduction to Kant's essay continues:

> ...Since the philosopher cannot presuppose any [conscious] individual purpose among men in their great drama, there is no other expedient for him except to try to see if he can discover a natural purpose in this idiotic course of things human. In keeping with this purpose, it might be possible to have a history with a definite natural plan for creatures who have no plan of their own.
>
> We wish to see if we can succeed in finding a clue to such a history; we leave it to Nature to produce the man capable of composing it. Thus Nature produced Kepler, who subjected, in an unexpected way, the eccentric paths of the planets to definite laws; and she produced Newton, who explained these laws by a universal natural cause.

Fig 2.3 Kant

Kant's essay seems to sense that the answer to his question must come from future research, and we can see that he is right: almost within a generation research starts converging on the clue: let us grant Jasper's and his immediate predecessors the discoverers of the clue..

The essay also asks for a demonstration of 'Nature's Secret Plan' and of evidence of a progression toward a 'perfect civil constitution'.

SEVENTH THESIS

> The problem of establishing a perfect civic constitution is dependent upon the problem of a lawful external relation among states and cannot be solved without a solution of the latter problem.

EIGHTH THESIS

> The history of mankind can be seen, in the large, as the realization of Nature's secret plan to bring forth a perfectly constituted state as the only condition in which the capacities of mankind can be fully developed, and also bring forth that external relation among states which is perfectly adequate to this end.

https://www.marxists.org/reference/subject/ethics/kant/universal-history.htm

mysteries. We ideas merge in an elegant unity: the solution to the Axial Age riddle in a larger cyclical system, the deduction of finite model from the idea of the transition from evolution to history, and the solution discovered to a famous challenge from the philosopher Kant. The field is now challenged by the intelligent design critics of Darwinian logic. Our analysis explodes the theistic account of the Old Testament even as it uncovers a far more complex indication of design, whose mystery we leave as systems analysis, not theistic mythology.

The current brands of evolutionary psychology attempting to apply natural selection to a reductionist version of human evolution has dominated biological thought since the era of Darwin but the problems with its core reasoning are well-known, if not well advertised to the larger public. Our strategy in the study of the 'macro' evolution in world history is to create a kind of systems model of its action. But this model requires an understanding of processes beyond the standard causal reasoning of normal science with a consideration of ethical and aesthetic issues that have no basis in conventional mechanistic accounts. To this must be added the essential distinction between the action of a system and the agents inside of that system.

Figs. 2.4 Champollion

This set of ideas will transform our analysis of the historical but it will in the process make it intelligible for the first time. This pattern of cyclicity is in turn related to a set of initial transitions and at a stroke we have the clue to both the enigma of the Axial Age and the rise of modernity.

2.2 From Fisher's Lament to Kant's Challenge

Fig. 2.5 Rosetta Stone

If we enquire into 'what runs history', into the possibility of any pattern, structure or law, we are left to examine the rush of statistics and wonder if it is sufficient to account for the chronicles of kings and commoners, the flowering of civilizations, and the evolution of religious forms. We are entering the forbidden zone, large-scale historical patterns, and have to deal with a considerable dialectic. Thus, the historian H. A. L. Fisher, in one of the most quoted statements of modern historiography

A Riddle Resolved

Big Histories, Universal Histories

Our account proceeds from causal Big History to Universal History, the evolution of freedom, and we can set up the starting point of 'Big History' as a backdrop to our search for a 'Universal History'. The idea of Big History, history since the Big Bang, is developed, for example, by David Christian in his *Maps of Time*, and this is also appropriate for our tale. Ironically this absolute beginning may in fact turn out to be another relative start, since Big Bang theories may or may not establish absolute starting points, and in any case this forces on us the question of evolution in its most general cosmic context. The connection between the two, self-evident in the eonic effect, is indicated by Christian de Duve in his *Vital Dust*, where the emergence or evolution of the human will in relation to values becomes a challenge to purely reductionist views. Reductionist science simply disregards the demand for any account of this aspect of evolution.

> **The Goldilocks Enigma** Paul Davies in *The Goldilocks Enigma* asks, Why does the universe seem so well-suited to life? Is this not really the answer to its own question: the transition from Big History to Universal History is effected by this 'fine-tuning' emerging in the Big Bang itself. Physics itself, although physicists are reluctant to admit it, gives us a hint of the mechanism beyond natural selection. This insight has been confused by metaphysical design arguments. But the empirical basis for a consideration of evolutionary directionality, beyond random evolution, is there.

Because of its double aspect, the idea of Big History stages a dramatic, almost drastic contrast of scales, the unimaginable vistas of deep time, next to the evanescent moment of man's emergence into Civilization, and our detectable 'evolutionary moments' at the level of centuries. We should peg our depiction of the latest with the earliest.

insists that there is no meaningful structure to be found in the randomness of historical process:

> Men wiser and more learned than I have discerned in history a plot, a rhythm, a predetermined pattern. These harmonies are concealed from me. I can see only one emergency following upon another as wave follows

Fig. 2.6 *The Course of Empire Destruction*

> upon wave, only one great fact with respect to which, since it is unique, there can be no generalizations; only one safe rule for the historian: that he should recognize in the development of human destinies the play of the contingent and the unforeseen.[6]

Increased perspective in the rising tide of historical data forces us to consider the counter-evidence to Fisher's Lament. Undoubtedly the influence of Darwinism is at work in Fisher's despairing rejection of any 'idea of a universal history'. The exclamations from the 'iron cage' of scientism in the wake of the seeming triumph of universal causal science seem to conclude the matter. But the triumph would seem premature, and the reign of Darwinian assumptions short-lived. History remains to be discovered. We live in a unique period of history, one in which the record of archaeology has begun to speak. Foreshortened perspectives of the historical have proven misleading.

Even as Fisher wrote, the record of civilization was crossing a minimum threshold of five thousand years to show a pattern of the type Fisher could not find emerging in fixer. We find an answer to the issue of historical rhythm, answers, but what was the question? Confusion over the nature of

A Riddle Resolved

The Non-random: The source of a model (From LFM)

The question of evolutionary theory haunts the efforts to create a science of the historical. No such science of history exists: there the issue of human nature defies easy analysis with man inside the problem. But we can suspect such a science in the evidence of the non-random.

The term 'non-random' is the safest way to refer to a perceived dynamic we don't understand fully. Crusoe who sees a footprint in the sand shows an example. But ripples in the sand would also qualify. The first case 'designed' the second causal, and marginally 'non-random. World history combines both and we see 'ripples' or cycles that carry interior cultural 'designs'.

The question of this 'pattern' arises in the context suggested in a famous essay on history to find a pattern of universal history. Although fuzzy, with a dynamics that is ambiguous, our pattern fits the bill, and there are probably no other solutions to 'Kant's Challenge' than the one we uncover using a frequency analysis, over the range of world history for which we have data at the level of centuries, or less. That's it: pattern shown. But what does it mean?

The 'non-random' is any evidence of a patterned process or dynamic. This could be causal, meta-causal or teleological or even designed. The evidence of world history contains a surprise: non-random patterning that exposes a 'macro' effect and provides a way beyond the dilemma, with a special kind of model. We see a rough solution to both questions: history and evolution (human) are Janus-faced. Our study will work with a model of this phenomenon and is like a passage through an enchanted woods. But it leaves a complex mystery that we must approach with care. Consider the phenomenon of the Axial Age: the data suggests the meta-causal, that is, some kind of semi-causal process that operates on some different level, acting on history rather than in history. The distinctions of causal event, machine and engine evokes something similar. Or perhaps the meta-causal process is really a teleological one, that gives evidence of directionality in the form of a cyclical process. This in turn hints at a distinction of the Kantian noumenal or phenomenal. This is the reason we limit ourselves to a broad set of outlines. The full model remains somewhat ambiguous.

historiography and historical theory makes the idea of a science of history or interpretation in terms of 'historical laws' uncertain.

Fisher's lament, with a tragic flourish, was perhaps a pessimistic or proto-postmodernist reaction to the horrors of the First World War, and the shock this created in the hopes of so many in automatic progress. His evocative statement was made in the wake of nineteenth century ideas of unlimited progress, and earlier ideas of universal history and is an indirect expression of the view that there is no discoverable historical pattern or direction. Beside it lie the many attempts to challenge the great philosophies of history that arose in the Enlightenment passing into the phase of German Idealism, then followed by efforts to approach its study scientifically, or the reaction to philosophies of history in the various forms of historicism, beginning with Herder. The current postmodern critique, the 'incredulity' toward metanarratives, joins the list of the skeptical judgments.

Fig. 2.7 The interaction of free agency and determination is visible in the simplest case of the computer mouse. Classic physics is surrounded by a fog of variants to determinism. One is Kant's idea of a 'causality of freedom'

Fisher's lament bundles together four, or more, quite separate concepts, that of rhythm, plot, pattern, and predetermination that do not necessarily stand or fall together. That historical patterned emergence can also be a series of chaotic 'emergencies', such as the French Revolution, is still another crisscross of meaning. A rhythm need have no plot, and a dramatic improvisation might show little or no predetermination, and yet operate under the constraint of a conditioned future.

The hold of Fisher's lament on many quotation-mongers and historical handwringers, as the magic sword to slay the dragon of macrohistory, is also a testimony to the difficulties of the project of Universal History, and its cousin, the attempt to find laws of history. Although the trend of current historical thinking, in the afterglow of the 'positive challenges' of positivism, is against the perception of meaningful historical structure, the plain fact is that the rise of the philosophy of history is a foundational moment for secularism and the understanding of modernity. If anything the rise Darwinian scientism is regressive.

As we discover the data of the Axial Age we begin to see that the kind

A Riddle Resolved

A Science of History? What is the relation of our method to Kant's actual system? There is a direct one in his so-called Third Antinomy.

"Causality according to laws of nature is not the only kind of causality from which the phenomenon of the world can be derived. It is necessary, in order to explain them, to assume a causality through freedom." Its antithesis is: "There is no freedom: everything in the world takes place solely in accordance with laws of nature."

We confront the enigma of the thesis, that freedom generation and physical causality somehow are both the case. The dilemma is immediate from the periodization of our model, remembering that this is only an empirical discovery, not a deduction.

Kant's Third Antinomy is reflected in our pattern, but on such a large scale, and such a different mode, that we must proceed with caution. From the way we set up our model (for another purpose) we can see how the stream of history seems interrupted by a second different 'causal initialization' that has no continuous lead up or antecedents. Our transitions are formally analogous to the noumenon, but quite different. They stand in conjunction to the limits of historical representation.

A frequency deduction A system 'evolving freedom' cannot cause freedom directly, since the over-determination would be causally closed. But such a system cannot leave action alone, since under-determination would not evolve freedom. Therefore, to evolve freedom such a system might alternate between higher and lower degrees of freedom, in cycles of macro-action, and micro-action left to its own devices. All at once we see that this corresponds to the eonic pattern. Thus, for example, the Axial Age shows a higher degree of freedom, but under eonic determination, while the mideonic intervals show the potential for freedom without the action of the system, 'real freedom', or not. The frequency system might terminate at some point to allow the realization of this potential. At the end we will suspect that we are at the end of the eonic sequence since observing the eonic effect probably preempts its future action.

of pattern in question is actually quite visible to the naked eye in proximate antiquity (the period from the first millennium BCE). We can match Fisher's lament with a challenge from the philosopher Kant to find such a pattern of history. An additional clue to the whole question lies in a simple question and a paradox that it creates: Is there a science of history? This forces the simplest dilemma: if there is such a science, there can be no freedom. We might seek the resolution by asking if there is some 'causality' of freedom that should accompany its appearance. If so we must find some evidence of its evolution. The study of history theoretically has proven intractable but world history must somewhere show at least some hint of resolving this field of contradictions. In fact, as we examine world history once again with this in mind, we suddenly discover that theoretical derivation matches the empirical record. This question was the object of Karl Popper's strictures on what he called 'historicism', and Isaiah Berlin's discourse on 'historical inevitability'. But the original version of this thinking appears in the philosopher Kant, who proposes it as the gateway to the philosophy of history.

One of the deepest currents of modern thought, beside the rise of theories of evolution, lies in the heritage of the philosophy of history, whose existence is justified by default in the failure to find a 'science of history'. No use complaining that science has replaced philosophy or that Darwin explains everything. Our simple model with its discoverable mainline stages a lightweight transition through this terrain. Strictly speaking our model based on a stream and sequence contrast, but then in this chapter has annexed the ideas of 'causality and freedom' as an adjunct, which requires explanation in the imperfect match. It is also empirical and can't be used for complex secondary deductions, but we can manage a few hunches with our historical black box, and the embedded freedom sequence tweaks the issues very directly.

We have found a solution to the paradox of causal determinism and the emergence of freedom in history: we see a macro oscillator shifting gears in its dialectic of 'degrees of freedom'. Beautiful. Our analysis blends in with a classic theme of the philosophy of history seen in the Dialectic of the *Critique of Pure Reason*, with its discussion of the various antinomies of reason, the so-called Third Antinomy being the key to our historical logic.

As we move to examine theories of evolution we find the philosophy of history's seemingly outdated, almost archaic, charm resurfacing as

System Action, Free Action:
Determinism vs Creativity

Related to the issue of Kant's Challenge is the issue of creative history, and we need to set a distinction, before embarking in the next chapter on a study of world history. The data of history is confusing unless we distinguish a causal factor from free agency, AND be sure to keep the two together, in tandem. We have evoked Kant's Challenge, and we must distinguish historical dynamics from free will, since **both** are operating, and we can call this the distinction of a system and the free agents inside it. Think of a ship and its passengers: the action of the system, the ship, and the action of the passengers on board is a hybrid system of mechanics and free will. It is important to see that history is not determined: it shows many hybrid situations where behavior is partly determined and partly free in the creative action of individuals. This distinction of system and agents might seem confusing, but we already know all this: the simplest example of the many we encounter every day might be the 'system action' of a car, and the 'free action' of the driver. The point is that 'history' has a mind of its own, so to speak, and we are inside it operating with our agendas. But the two intersect. We need a looser version of the duality of causality and freedom: system action and free action. Free agency is not always 'free will'. You can be a free agent in an earthquake, but not free to do much of anything while it happens!

Some analogs The simplest example here is that of a driver in a vehicle. The situation shows the tandem action of a causal machine and a free agent, with our without free will, in control of that machine. Another example is that of an ocean liner and its passengers. Still another is a computer with a mouse, a clear tandem situation of 'system' (computer) and 'free agent', user with mouse.

a renewed challenge, and an obstacle to their completion. If a theory of evolution moves to enlarge its domain to include the whole, then it is forced to reckon with the self-reference of the thinker pondering his own evolution. No other grounds are required for the persistence of this mode. The idea of evolution is a feckless giant, and we should propose, in a gesture more than humor, a comeback of philosophical history, a nimble rascal, to leap and ride piggyback, wishing to direct traffic, to the consternation of proponents of post-philosophical science. Indeed, we should notice at once that the philosophy of history is itself a part of our universal evolution, as is the idea of evolution, that is, the evolution of the idea of evolution.

Displaced in the rise of the positive sciences by the idea of evolution, the philosophy of history becomes one of its first passengers. For the philosophy of history is the history of philosophy, and this shows the signature of its own (eonic) evolution. We can offer no real differentiation, then, of the two subjects, or any decisive means of marking the transition between boundaries of rival disciplines. If Darwinism is free of metaphysics, then let it be science. But we have seen that it fails three times, in the classic antinomies given from Kantian Dialectic.

The philosophy of history is born, reborn, at the dawn of modernity as a fellow traveler, becoming visible as early as the sixteenth century and finds its classic realization in the writings of the philosopher Immanuel Kant, in his essay *Idea For A Universal History from a Cosmopolitan Point of View*:

> Whatever concept one may hold, from a metaphysical point of view, concerning the freedom of the will, certainly its appearances, which are human actions, like every other natural event, are determined by universal laws. However obscure their causes, history, which is concerned with narrating these appearances, permits us to hope that if we attend to the play of freedom of the human will in the large, we may be able to discern a regular movement in it, and that what seems complex and chaotic in the single individual may be seen from the standpoint of the human race as a whole to be a steady and progressive though slow evolution of its original endowment.[8]

This hope is confirmed by the pattern we can exhibit, and we can derive the result from this paragraph. The inherent contradiction in this paragraph does indeed generate its own historical dynamic.

> To preview our results we can note that the solution to the enigma of the Axial Age and to Kant's Challenge will turn out to be the same problem!

A Riddle Resolved

And the eonic effect answers at once to the question asked. Kant's essay is constructed around a classic ambiguity on the one hand it seems to propose a solution to his own question and at the same time throw the question into the future, for an historian with greater perspective to discover an aim of nature in the chaos of historical happenstance. This is strangely apt: archaeology has begun to provide answers here in the wake of his query.

Fig. 2.8 From *The Holy Land*, David Roberts

Kant's essay is one of the few that have stumbled on the key concept of the 'causality of freedom'. This constellation of ideas is very hard to analyze but is useful in forcing us to bypass debates over free will by seeing it in the context of a larger system. And the 'evolution of free will' needs to be seen in the context of human evolution.

Beside this projection into the future of this wish to discover 'nature's secret plan', Kant also relates the issue to the idea of progress toward a 'perfect civil constitution'. Kant's essay seems almost perfectly tuned to our study, without realizing it, for our discovery of 'historical evolution', as we will see, throws light directly on both of these issues, exhibiting the reality of 'nature's secret plan' behind the emergence of civilization and more specifically the directionality in the development of civil government. As we proceed we will see the remarkable way that the eonic sequence demonstrates a law of progress, and of the concealed teleology behind the evolution of culture in world history. And the particular pattern of political development inside this progression will exhibit the way in which emergent democracy is bound up in the eonic effect itself.

> Kant's essay invokes issues beyond its cogent insight into free will: nature's secret plan, and the creation of a 'perfect civil constitution'. These questions are directly relevant to the issue of the Axial Age, but only in the larger system of such 'ages' or axis points. Our model scores a home run on all these questions.

Fig. 2.9 The Divided Kingdom

Axial Age Israel is the most striking case of 'spooky design' hallucinations. It offers a pattern of 'coincidence' that bursts beyond the framework of sociological mechanics. The 'disappearing kingdoms' drama of Judah and Israel and the last minute braiding with Persian Zoroastrianism is a feat no standard theory of any kind could explain. But the 'theistic' mythology that emerges is itself output of the 'system in action'! A new 'secular' interpretation of the data would actually be more remarkable than the Biblical version. We must proceed with caution, since the whole field is an invitation to error.

A Riddle Resolved

As we examine world history the data emerges clearly to resolve Kant's Challenge in unexpected fashion. We have the framework to proceed with an outline of history as the 'evolution of freedom', starting in the next

2.10, 11 Adam and Eve expulsed from Garden of Eden

chapter. The great irony here is that we will see Kant caught up most beguilingly in the very turning point that constitutes one aspect of his problem's solution. The answer needs just a bit more time and perspective. It is a beautiful prophecy and proof of the power of his system of critiques.

Within two centuries the necessary data is emerging for the first time to resolve Kant's Challenge in unexpected fashion. We can easily resolve the question of directionality, but not fully that of teleology. Directionality, seen in the evidence of past times, expresses the phenomenal representation of some inferred teleological process, whose outcome, or telos, however, is beyond observation, and in any case a timeless unknown with its foot in the future. Of this we can know nothing as our eonic system is seen, looking backwards, to have proceeded toward the present in the recursive approximations we see in the eonic sequence. And we isolated one theme of that progression as an 'evolution of freedom', as an empirical study, without committing ourselves to any generalization beyond our present. Our approach is indirect, and the reason is the danger of premature teleological metaphysics, which ends in limbo if we give it an answer without an ending, which requires some statement about the future and/or the eonic sequence. But that very caution is implied by Kant's essay.

The pattern of the eonic effect is not a philosophic solution to a problem, but an archaeological finding, partial in the sense that a shard of some lost whole is discovered empirically. Our pattern for all intents and purposes

Fig. 12 The Exile

answers the quest initiated by Kant, seen in the subtle wording of his remarkable formulation, *itself correlated with the pattern*, that we should *attend to the play of freedom of the human will in the large, to discern a regular movement in it*.

2.3 The Old Testament and the Axial Mystery

The phenomenon of the Axial Age is a paradoxical question already known to us under another guise. The Old Testament is a puzzle with tale about a Canaanite cultural region that undergoes a very mysterious historical transformation. In a secular age that tale is subject to equivocations of belief or skepticism. But the larger issue is the way the saga of 'Israel' (in quotation marks, the account is about a pair of kingdoms) echoes a larger effect stretching across Eurasia, in the centuries from about 1000/900 BCE to the period of the Exile and beyond, ca. 600 BCE. Once we see the connection the tale is confounded by a double-take of explanation: we seem to see a sociological effect behind a now very ancient epic account of a people called the 'Israelites'.

2.13 Nazareth, David Roberts

The clarity of the historical interval in question, in the sense of being a purported factual account of kingdoms confronting the Assyrian and then Persian empires, contrasts with the extended mythological opening starting with the book of Genesis. It is not until we decipher the larger phenomenon of the Axial Age that this data begins to make sense of a book that seems to have forgotten its deepest meaning. The issue is a riddle of two things in a state of overlap: the larger semi-mythological epic of the 'Abrahamic' source tale, recounting the interaction of a set of tribalisms with the greater Egyptian civilization, ending in the tale of Moses, and the tale of the conquest of Canaan, now thought to be entirely mythological. What is the solution to this mystery.

What we are really seeing is almost more remarkable: the clear stages in the formation of a monotheistic proto-religion, one still entangled in cultural in the specifics of its time and place, and moving slowly out into a larger environment. But the issue is ambiguous in many ways: the larger history of the period shows, strangely, a parallel phenomenon in the Persian world adjacent to and then the conqueror of the great source civilizational zone of Mesopotamia. There again we see an emergent monotheism, with a collation of Abraham/Moses in the figure of Zarathustra, whose legacy will be a second monotheistic 'branding', this time encrusted with its own Indo-European cultural guises. Then, remarkably, these two brands of cultural monotheism meet during the Exile period. It is not always clear what the resulting blend owes to each source, although clearly the Hebraic version moves more clearly toward an Occidental realization as Christianity.

In an era proclaiming itself as 'secular' the correct explanation of this record of emergent monotheisms is liable to be dismissed in an attempt to expose the religious mythologies of two separated cultural streams. One problem is the lack of correct historical information about the cultures in question. The Old Testament is treacherous: it overflows with token historical facticity, only to drift suddenly into entirely made up additions to the account.

Fig. 2.14 Zoroaster

Even as we explore the riddle of the Old Testament history of Israel we discover a similar emergent monotheism in the legacy of Persia, and this, as history records it is blended with the Israelite version as the time of the exile. It is interesting that the roots of millennial conceptions in their current form emerged from the ideas of Zarathustra, in the second Millennium BC, passed through the vehicle of the Persian Empire into the parallel world of emerging Judaism during the period of the Exile and thence into Christianity and Islam. By this reckoning our crisis is quite ancient indeed, as recycled eschatology. It is difficult to reconstruct the exact relationship of Zoroastrianism and the Hebraic monotheism, although the *Book of Daniel* shows the clear footprints leading back to the era of the Persian Empire in the time of Cyrus the Great.[1]

Our sense of universal history springs from the Old Testament

---

[1] Norman Cohn, *Cosmos and Chaos and the World to Come* (New Haven: Yale University Press, 1993), *In Pursuit of the Millennium* (New York: Oxford, 1970), Theodore Olson, *Millennialism, Utopianism, and Progress* (Toronto: University of Toronto, 1982). Peter Clark, *Zoroastrianism*, Brighton: Sussex Academic Press, 1998. Albert Schweitzer, *The Quest for the Historical Jesus* (New York: Macmillan, 1948).

A Riddle Resolved

epic. But this is a complex hybrid of multiple origins. The blend of indigenous Judaic monotheism, as it emerged from its Canaanite, thence Egyptian and Mesopotamian traditions, along with the themes of Iranian dualism and eschatological messianism during the period of the Exile and after, resurfacing strongly during the Qumranic period near the birth of Christianity, is one of the most confusing overlays of the period of cultural advance and integration that occurred with a center of gravity ca. -600, thence to generate the pillars of a great constellation of traditions. This complex parallel emergence and interactive blending constitutes one of the central mysteries of the western religious tradition.

2.15 Abraham

That the record of the period of Exile given in the Old Testament should have preserved the forgotten connection of eschatological ideas with the parallel Zoroastrianism in the world of the Persian Empire is a piece of a greater puzzle. It is the period ca. -600, plus and minus, that is in fact our subject, for it is this era that is the rough center of gravity of a great transformation, known as the Axial Age.

It is the era of the birth of the great religions *in concert* at the fountainhead of the traditions of classical antiquity. The process transcends the phenomenon of religion and we see that the synchronous effect applies as well to the polytheistic Greece in the period of the Ionian Enlightenment. The seeds of modern secular culture are there sown at the same time, there is no clear differentiation. The Old Testament conceals a riddle, but cannot do justice to its own discovery of the Axial Age. Its perspective is too localized.

> **The Birth of Universal History** The Biblical tradition gives testimony to the birth of ideas of universal, or progressive history, against the backdrop of cyclical myths, and this was influenced by Zoroastrianism. The irony that this linear, eschatological view of history should emerge in the mysterious moment of the so-called Axial Age, whose cyclical interpretation we will discover, and which will drive us to see their synthesis, the cyclical driving the linear, in the eonic effect.[2]

---

2 As Norman Cohn notes in *Cosmos, Chaos, and The World To Come* (New Haven: Yale

The myths of the Old Testament require a new understanding in the wake of the findings of Biblical Criticism, and the phenomenon of the Axial Age. We need to recast our understanding of the remarkable significance and context of the Old Testament. It is pointing indirectly to a great historical transition, in the evolution of religion itself toward a new form of monotheism. But that transformation is larger than the phenomenon of religion.

Even secular philosophy finds itself unable to do justice to this seminal epic at the dawn of middle antiquity. It is important to consider how little accurate information we have for this period. By comparison the histories of the Greek period are rich in data. We could not reliably speak of the historical existence of Abraham, Moses, the Exodus, or any of the other details of a history rendered into an ideological collation in the generation before the Exile.

> **The Bible Unearthed** A renewed sense of the extraordinary significance of the Old Testament leaves us with a question, What is the Bible recording? Theistic historicism or an Axial transformation? The natural division into three sections, the Torah, the Prophets, and the post-exilic writings of the period Ezra and Nehemiah, gives the clue: the prophetic period straddles the Axial interval and this, as we will see, is period of transition to a new era, leading to its conclusion at a point of 'divide', ca. -600, in its enigmatic synchrony with Greek, Indian, Chinese, and other parallels. We can decipher this transition by comparison with its isomorphic instances, as in the emergence of Classical Greece from the Greek Archaic. The Bible comes into existence and begins to crystallize in the generation of the Great Reformation of Josiah at the conclusion of its Axial transition.[3]

Seen rightly, the Old Testament's *core account*, the rough interval from -900 to the Exile, unwittingly records an incident in the Axial Age. The puzzle of continuity and discontinuity perplexed the redactors of the Judaic corpus

---

University Press, 1993, p. 227), "Until around 1500 BC peoples as diverse as Egyptians, Sumerians, Babylonians, Indo-Iranians, and their Indian and Iranian descendants, Canaanites, pre-exilic Israelites, were all agreed that in the beginning the world had been organized, set in order, by a god or by several gods, and that in essentials it was immutable…Some time around 1500 and 1200 BC Zoroaster broke out of that static yet anxious world-view. He did so by reinterpreting, radically, the Iranian version of the combat myth."

3 Israel Finkelstein & Neil Silberman, *The Bible Unearthed* (New York: The Free Press, 2001), William Dever, *Who Were The Israelites and Where Did They Come From?* (Grand Rapids, Michigan: Eerdmans, 2003).

A Riddle Resolved

who attempted to seek the sources of their suddenly appearing tradition in earlier figures. Yet the sagas of Abraham and Moses, if historical, clearly precede the crucial phase. One irony of our enquiry will be to inherit the true beauty of the Old Testament in a secular interpretation.[4]

This period seems the source, as an age of 'revelation', of our sense of the sacred. Yet we can now see that the Zoroastrian, Abrahamic, and other sources *precede* this period, whose relative transformation of outstanding cultural streams seems to generate the illusion of an absolute or transcendental source. This is a challenge to our idea of an age of Revelation. Further, Christianity and Islam arise much later, but seem to look backward to this period, whose actual core shows something quite different, the history of a Canaanite culture zone, 'Israel/Judah', whose religious traditions suddenly transform into a monotheistic vehicle, as it sows the seeds of the religions to come. An almost identical phenomenon, at this high level of abstraction, is visible in India, and in a comparable time frame. In fact this entire period was extraordinary in its generation, and all at once, of new cultural traditions. The complexity of this picture requires a new type of historical model.

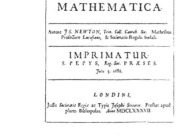

Fig. 2.16, 7 Newton and *Principia*

The Evolution of Religion? The Old Testament records a paradox: monotheism seems to begin with an 'Abraham', yet also seems to come into existence in the Axial interval. This problem of relative transformation is a prime candidate for analysis using our eonic model. The 'evolution' of religion in the emergence of civilization is a complex overlay of two processes, macro and micro. The micro aspect develops at all times, while the macro is expressed in a larger discontinuous series. The intersection of the two is what leads to the remarkable florescence we see in the Israelite monotheism that surges outward, like an amplified signal, in the wake of the Axial interval. One and the same effect, and one and the same timing, is

---

4 As Wellhausen suspected, it would seem that it was the period of the prophets that represents the real transformation that generates the emergence of monotheism. Cf. also, Giovanni Garbini, *History and Ideology in Ancient Israel* (London: SCM, 1988).

visible in the emergence of the parallel Axial Buddhism in India.

With the increase of modern historical knowledge this strange phenomenon of synchronous parallelism has become an enigma replacing a myth, in the process casting the Occidental myths of revelation in a most ironic light. This constellation of creative individuals generates a new age of history, and leads us into causal perplexity before such a complex temporal

Fig. 2.18 Early Factory

correlation over independent regions of so many effects. It is a phenomenon of *Gaian* proportions, yet we see only a series of outcomes, never the dynamic behind them. There is nothing simple about it, for while it is true that the Old Testament demonstrates the appearance of Biblical prophets in this period, the effect has nothing as such to do with prophets. Prophets existed before, but none quite like this unique series in their anticipations of a new world to come.

From its archetypal roots, the eschatological idea forever resurfaces, as evidenced in the versions of early modernism, as they influenced, for example, the German and English Civil Wars, Hegel, and Marx. The eschatological nexus moves between its twin realizations, the slow, and the fast, the one conservative dangling the carrot of hope, the other radical, pedal to the floor acceleration and social tumult. The 'end times' are the grounds for the last revolution, or else the 'end of history' is the rationale for the end of revolutions. It is no accident that much contemporary social criticism attempted to expose the fast version embedded in leftist communism, looking the other way at the slow version granted the weight of religious tradition.

A Riddle Resolved

The eschatological idea echoes throughout history, reaching the modern world in its inverted secular forms, such as the Hegelian 'end of history' showing the connection between state and transcendence in direct fashion. This thinking echoes the question posed by the philosopher Kant in his classic essay *Idea For A Universal History*, discussed at the start of this chapter. Our secular Zarathustras live in the acceleration of history, the exponential curve as myth. Francis Fukuyama finds, in *The End of History and the Last Man*, that we have reached a political final state, the end of world-historical political evolution in the form of the liberal state. If this is true, it should better be called the Beginning of History, the real New Age, if its creature could reach future history as a New Man. But the point is rather that in the perception of Hegel the evolution of freedom visible in the realizations of modern democracy tokens a New Axial transformation of the worlds inherited from antiquity. Finally, in the vault of time, the scale of the historical passes to the moment of Earth time and the evolution of life, thence to embrace the Big Bang and even, in new crypto-Zoroastrian theories of physics, a final relativistic Omega Point of converging world-lines at the "end of time".[5]

Fig. 2.19 *Declaration of the Rights of Man*

2.4 Decoding Modernity

Against the backdrop of world history the rise of the modern must constitute one of the most explosive turning points since the beginning of higher civilization, or even the onset of the Neolithic. In the three centuries after 1500 beginning with the Protestant Reformation and the parallel Scientific Revolution an entirely new form of civilization has arisen, set to transform the entire planet via globalization. Such a massive transformation demands an explanation on the scale of evolution itself, and shows a remarkable discontinuity against the backdrop of medievalism. But this issue has been confused by debates over traditionalism or medievalism. It requires a larger context for a solution to the riddle.[6]

---

5 Francis Fukuyama, *The End of History and the Last Man* (New York: The Free Press, 1992).
6 Jacques Barzun, *From Dawn to Decadence: 1500 to the Present* (New York: HarperCollins, 2000).

The sudden explosion of modernity is an empirical given of world history. And yet a sense of crisis now haunts the idea of the modern. Indeed, a renewed challenge to the meaning of secularism in a resurgence of religious traditionalism seems to threaten the legacy of the Enlightenment. There is even the invention of a spurious 'postmodern' age to replace the modern. These gestures might betray the agenda of reactionaries, but demand a reckoning of modernity in terms of world history as a whole. There can be no replacement of modernity with an ad hoc postmodern concoction. The result would be decline, not advance. The sudden explosion of the modern might well show 'action and reaction', with a waning of the original impulse. Yet defenders of modernity seem ill-equipped for the task of defending its significance against its critics, or meeting the crisis that threatens its realization and future. What is the source of this sudden chaotification?

The question confronts us, What is the significance of modernity, and how can we understand its sudden transformation of world history?

> **What is modernity?** We are left with the ambiguity of what we call the modern, next to the equal confusion over the meaning of secularism.
>
> **Is there a postmodern age?** One of the most radical attacks on modernity is the gesture to posit a 'postmodern' age. But this idea suffers a curious contradiction, and expresses an agenda that is ambiguously reactionary. Postmodernists have wished to 'deconstruct' grand narratives, but we might as well wish to deconstruct the flat histories that are the result.

In one sense, the crisis is real enough. Environmental catastrophe looms, as the Age of Oil seems destined to a swift conclusion. As if to summon the spectre of Marx all over again, the Industrial Revolution itself seems under siege as a Faustian gamble, the automatic dynamism of modern capitalism looms as a monster out of control. A postmodern gloom seems to have settled on the prospects of the new age spawned in the centuries from the Reformation to the Enlightenment. But the modern is far larger than its economic contradictions, which have no pre-modern solutions. We seem to confuse economic dynamics with the fact of modernity as an already irreversible stage of history.

Our situation is not helped by the incoherence in our views of history. Here the influence of evolutionary thinking next to the economic interpretation of history has blinded us to any sense of universal history. The result is a kind of Darwinian economic fundamentalism resulting in a reductionist inability to grasp even the significance of secularism, or to see the complexity of innovations to which we cannot do justice beyond the questions of technology and the Industrial Revolution. The rise of the modern is a puzzle in itself, an almost evolutionary break in the continuity of world history. Exploding in the sixteenth century with the Reformation and the incipient rebirth of the Scientific Revolution, the early modern ignited a transition to a new phase of human culture, and by the eighteenth century the foundations of an entire new era in world history had been laid, graduating in the climactic moment of the Enlightenment, the French and American Revolutions, and the onset of the Industrial era. And this is the historical transformation that has produced so-called secularism, and its collision with religious traditionalism.

There is an irony here: this phenomenon of sudden discontinuity is not unique and resembles the seminal moment of the foundation of our traditions. We can see clearly that a moment of great discontinuity, the onset of classical antiquity, was the source of the great religions as we know them now. But also, ironically, of the very secularism that now seems to challenge these traditions. It is altogether strange, and yet surely significant, that the age of the *Upanishads*, and that of the Israelites in the period of the Prophets, should occur in rough simultaneity, and gestate from the Indic direction the great religion of Buddhism, while in the case of Israel a reaction to polytheism should generate a new type of monotheism destined to characterize three subsequent religions of Judaism, Christianity, and Islam. We must pursue the investigation to the end, to find in the parallel age of Greece the seeds of modernity itself.

It is an odd pairing of opposites to see the parallel emergence of two world religions, of such different character. It is obvious that what we consider to be a secular age is a reaction to this legacy of the religions inherited from antiquity. But it is a reaction to their medieval construction. The period of their birth was something quite different. And these religious formations in turn were a reaction to the religions of their time. We should note that the rise of the secular is not so much a reaction against religion, as its transformation, visible in the Protestant Reformation. The distinction between 'sacred' and 'secular' is misleading. We seem to detect a cyclical phenomenon. And, the enlarging scope of our historical vista is starting to show us eras of religion far earlier than what we take as religious tradition. Beyond even the world of Egypt and Sumer we can observe the archaeological remains of temples already ancient by the time of the first Sumerian cities. We can begin to see that organized religion was already ancient by the time of the first Pharaohs, and that temple complexes were already in existence in the millennia before the rise of the first great technological civilizations of Sumer and Egypt.

Fig. 2.20 Huns in Italy

It is more than whimsical to cite a cyclical metaphor in a progression of epochs, for it will challenge us to consider the history of the many mythologies of cyclical history, and this in counterpoint to some reckoning of the idea of progress, the clue in fact to its reality. The trick is to reconcile so-called linear and cyclical views of history into a higher unity. The idea of progress has fallen on hard times, and in a postmodern period it is almost an idea in exile, and yet its significance for the rise of modernity is crucial, and its emergence in the early modern was as a challenge to the dominance of antiquity in the minds of those who began to see that what they called the 'modern' period was starting to outstrip the achievements of Greece and Rome. The ideological character of the idea of progress, and its degeneration into a form of economic propaganda, is a later development. The idea of progress was a great challenge to the myths of cyclical history, but there is an irony here, that the cyclical and progressive views of history might be reconciled in a fashion that actually demonstrates the progressive character of world history. Already as a first impression we have seen a series of discontinuities express the timing of a series of advances or reborn eras in world history, among them the rise of modernity. The riddle of linear progress is ironically resolved by seeing its cyclical aspect, an idea to confound cyclical myth-mongers.

The idea of progress is rejected by biologists in the discussion of evolution, and this has become one of the central dogmas of Darwinism, but at the very least the idea serves an essential function in our understanding of history, whatever the case with biology. Can we really look at the spectacle of emerging civilizations as a stasis of undeveloping entities? Clearly the notion that things are somehow in a process of development and complexification is indispensable in the attempt to chronicle man's historical emergence from the Paleolithic. We need a new way to look at the idea of progress, to see at once its ideological abuses, and its essential rightness or inevitability in any understanding of evolution. Part of the confusion lies in the obvious way in which what might be seen as periods of advance, are in clear contrast to the longer intervals, all too visible in history, of what might almost seem retrograde motion.

In fact, prior to the archaeological revolution of the nineteenth century, the Western view of world history consisted of the tale of classical civilizations beginning with the Classical Greeks, and the saga of the Old Testament, followed by the story of Roman turning into an empire, which endured for many centuries and then declined into a medievalism whose total historical interval outstripped all else, and dominated the historical portrait until the quite recent rise of the modern. This overall perspective was not conducive to clarifying the demonstration of progress in history. As we move backwards, a strange perception arises. The same constellation of advance, then a 'medieval' stasis, is visible in an earlier cycle, beginning with the surge of higher civilization at the end of the fourth millennium, in Sumer and Egypt, followed by the less seminal centuries enclosed by its beginning, that finally fades away into the decline preceding the rise of a new era at the time of the classical Greeks.

2.5 Decline and Fall

This brings us to the dynamical mystery of civilizations, their apparent rise and decline, and the misleading way in which a postmodern perspective has become a version of declinism. Modernity is barely underway, and yet a version of leftist or religious ideology has declared the 'age of modernity' to be finished. It is significant that the term 'postmodern' appears, before its appropriation by a cultic wing of the modern left, in the historian Toynbee. And next to Toynbee we have the figure of Spengler whose 'postmodernism before the fact' defines very clearly the genesis of the postmodern reaction to modernism. This in turn shows the clear influence of the philosopher Nietzsche whose attack on modern liberal civilization is one of the pivot points of the anti-modern reaction. The thinking of Toynbee and Spengler has proven strangely influential despite the many critical exposés of the limitations of their historical models.[7]

The idea of the 'civilization' is central to the thinking of Toynbee and Spengler whose works constructed a kind of botanical classification of the various specimens of such, and the result has been a rigidification of the concept as some kind of dynamical entity, or even as an expression of the organismic. And this in turn leads to some notion of the lifespan of a civilization, resulting in the predictable onset of its decline. The great exemplar is the 'decline and fall of the Roman Empire', which becomes by analogy the misleading template for editorializing the fall of modernity. And this declinism has become the warning cry of many 'Spenglerians in spite of themselves' who are nervous that the 'modern civilization' is about to enter the final stages of Rome's later empire. There is something amiss in this reasoning. The modern world is a mere centuries from its dramatic initials incidents, such as the Enlightenment. It would seem a desperate shortening of a potential future for this to be already in decline. Between the onset of the Roman Republic and the final decline of its empire is an interval of a thousand years.

Toynbee seems to wish for a new manifestation of traditionalism, Spengler a renewed barbarism in the aesthetics of Nietzsche. There is something confused about this legacy of Toynbee and Spengler, and it becomes important to try and come to an understanding of the limits of their analyses of world history, with their concealed cyclical perspective. The rise and fall of civilizations is not a difficult concept to document, up to a point, in the chronicle of civilization, but something is awry in the methodology of these two thinkers. We can see the problem perhaps in the way Spengler concocts a 'Faustian civilization' for the West, beginning in the year 1000, and now reaching its final stages. Can this be right? The arbitrary start at the moment of the first millennium, the depiction of the rise of the modern period and the Enlightenment as somehow the approaching decline, and the final 'decline of the West' trumpeted at the beginning of the twentieth century leaves one to ask if the concept of 'civilization' is really the right one

---

7 Oswald Spengler, *The Decline of the West* (New York: Knopf, 1926), Arnold Toynbee, *A Study of History* (New York: Oxford, 1957), abridgement by D.C. Somervell.

for the study of the historical dynamics of the modern 'west'. The civilization, as a rubric is directly intuitive as a descriptive device, but the moment we begin to make assumptions about its 'evolution' in some fashion, we seem to be on less certain grounds. There is a much simpler pattern of civilizations than that of their rise and fall. We see a progression of eras beginning with the rise of higher civilization in a system that transcends civilizations and seems to generate Civilization, in a process of localization and globalization.

The gloom of Spengler is in one way understandable, composing the elements of his immense tome against the backdrop of the First World War whose unexpected savagery left the idea of progress shattered in the minds of a whole generation. It seemed as if the hopes and expectations of modernity had been betrayed by a regression. And there was worse to come. The unimaginable, like a cusp in history, was soon to emerge in the convulsion of Nazism and the Holocaust. It was, and is, hard for many to even consider the idea of progress again after such an unprecedented outbreak of the demonic. And yet the very tone of Spengler's perspective, with its implicit Nietzschean embrace of wars to come and to be unparalleled in their virulence, is itself the self-destructive omen, the curious prophecy of the psychosis that seemed to overtake the 'West'.

And yet the intervening years did not really show the decline of the West. Perhaps it has demonstrated globalization beyond the vehicles of the early modern, or the limits of imperialism in these incipient champions of the modern. But this might be progress, not decline. From the First to the even more cataclysmic Second World War and beyond the fate of this 'west' was one of triumph and recovery, and a second act of the realization of modernity. And the very notion of the 'West' began to yield to the globalization of its idea, and the creation of a new and larger oikoumene. For the idea of the modern competes with the idea of the civilization, as a term of periodization, and has no geographical or cultural bounds. We become suspicious that the idea of some 'western civilization', with its inherent Eurocentrism, has missed the point. There is a flaw therefore in the idea of the 'civilization' as the basic unit of analysis, in some organismic metaphor of its life. For the larger direction of history has shown the supposed civilization of the 'west' to be an appropriate stepping stone toward a larger sphere of modernity, which is more than a civilization.

The American Empire? The theme of leftist critique of American imperialism has recently seen a revival of the declinist genre applied to the United States of America. In *Nemesis*, for example, the author sees the analog of the lost of the Roman Republic in the American democratic system. This is a somewhat more relevant comparison than to the fall of the Roman Empire, but the very nature of this periodization could be misleading. In any case, the challenge to imperialism is not the same as the decline and fall of a civilization.[8]

The study of history would seem to require a larger concept than that of the civilization. The issue appears to be not the lifetime of a culture, but the interval of transition to a new era, and the spread by diffusion of its idea, in the creation of an oikoumene. Once we adopt this altered perspective, many examples come to light. The lifespan of Greek civilization is very long, stretching from almost the Neolithic to modernity, and undergoes many changes in the form of its culture. But this is not necessarily the right concept of its history. Rather we see that this stream of historical culture has given birth to a whole series of significant moments, of lesser duration. The great classical era of Greece, which produced a turning point in world history, was merely an interval of short duration, several centuries, in a mysterious flowering of culture, one that, just as with modernity, produced by diffusion a new and larger oikoumene in a process of incipient globalization.

The brief era of the flowering of Classical Greece is one of the most remarkable in world history, and behind a disguise closely resembles the rise of the modern. It is in fact the birthplace, however inchoate, of the secular. The remarkable thing about this was the speed, and brevity, of the transformation. Between the eighth and fourth century BCE the entire spectacle of the Classical Greeks opens and closes, leaving behind an achievement whose immensity remains with us to this day as one of the foundational moments of Western, we should say, world civilization. We cast about for some means to explain this apparition in world history, but are left with an absence of clues of the sociological variety. We assign causes to antecedents, but if we examine early Greece emerging from its Dark Age we are left empty-handed as to causal explanation. What sociological factors could we list that might explicate this spectacular phenomenon? Probably none. We need a new perspective altogether.

In our search for the causes of the Greek achievement, sometimes

---

8 Chalmers Johnson, *Nemesis: The Last Days Of The American Republic* (New York: Henry Holt, 2006).

A Riddle Resolved

Fig. 2.21 Biblical Lands

Frontier Effects

At first the Axial Age zones of effect make no sense, and many key effects also occur outside the Axial regions. But the phenomenon needs to be understood in a larger context, and doesn't depend on there being related effects outside its fields of action. The whole effect (which needs a new name) follows a fairly simple set of rules. One is a frontier effect. Why don't we see any action in core Sumer or Egypt in this period?! A simple frontier effect is at work: we see that Israel and Greece are both frontiers in the larger system, and that the series of hotspots never repeats itself. Why would the 'Axial transformation' bypass Egypt and start over in a small area of Canaan? Our model explains this easily.

The Frontier Effect A key property of our eonic pattern is the 'acorn or frontier effect'. Note that something global is occurring starting in a series of local areas. But the sequence restarts in a new place each time, like an acorn, just at the frontier of its predecessor. The world of Canaan, spawning 'Israel', does not look like a frontier now, but in the era of the mythical Abraham it certainly was, and we even have a 'pioneer' story about his leaving the city of Ur in a prime diffusion source, the world of prior Sumer. Greece and Rome in the Axial period were definitely still frontier areas, relative to the by then ancient world of Egypt and Mesopotamia. Each of our transitions creates a hotspot, then expands to create a new civilization, better, oikoumene. Cultural acorns sprout in this field, and then at the next cycle one of them becomes a new transition. Note how our sequence is generating 'evolution in the large' via local hotspots, 'short term evolution in the small'. We must study the diffusion fields of our turning points. From WHEE called the 'Greek Miracle', we are left with the impression of something uncaused in its suddenness of emergence, and also with the unsettling data of synchronous phenomena in several places at the same time. Even as the Greeks in a strange spontaneity emerged from their Archaic period to a moment of greatness, nearby, and in a strange simultaneity, the drama of the Israelites was playing itself out, as the epic of a Canaanite people, again almost a frontier culture, who inexplicably entered the world stage with the creation of a new monotheistic conception of religion, and a great literature, parallel to the Greek, documenting the stages of the emergence of this challenge to polytheism, and the religious heritage of civilization, outstanding since the Neolithic. We are coming to one of the most significant discoveries of modern historiography, that of the Axial Age.

2.6 Discovery of The Axial Age

Our search for causes is confronted with the phenomenon of the so-called Axial Age, a term invented by the philosopher Karl Jaspers who collated a whole series of observations of this phenomenon, as it came to be discovered in the nineteenth century. The discovery of the Axial Age is one of the great episodes in the more general drama of the archaeological revolution, whose most notable incident is perhaps the discovery of the Rosetta Stone by the army of Napoleon in its invasion of Egypt. The sudden opening to the mystery of ancient Egypt in the decipherment of its ancient hieroglyphics heralded the massive new findings of the nineteenth century. The at first less spectacular but in many ways as significant discovery of the Axial Age did not impinge on public consciousness until much later, and in fact has still not done so. From his *The Origin and Goal of History*, we have Karl Jaspers' observation:

> The most extraordinary events are concentrated in this period. Confucius and Lao-tse were living in China, all the schools of Chinese philosophy came into being, including those of Mo-ti, Chuang-tse, Lieh-tsu and a host of others; India produced the Upanishads and Buddha and, like China, ran the whole gamut of philosophical possibilities down to skepticism, to materialism, sophism and nihilism; in Iran Zarathustra taught a challenging view of the world as a struggle between good and evil; in Palestine the prophets made their appearance, from Elijah, by way of Isaiah and Jeremiah to Deutero-Isaiah; Greece witnessed the appearance of Homer, of the Philosophers—Parmenides, Heraclitus and Plato—of the tragedians, Thucydides and Archimedes. Everything

implied by these names developed during these few centuries almost simultaneously in China, India, and the West, without any one of these regions knowing of the others.[9]

Our perception of the suddenness of the Greek transformation, and the parallel emergence of the prophetic age of the Israelites now finds its explanation, or rather a larger question in search of an explanation, in the realization that an entire spectrum of cultures across Eurasia in the period, as Jaspers depicts it, from -800 to -200.

Here simple periodization uncovers something spectacular, however we are to interpret the result. And yet this discovery has been almost orphaned by an inability to properly grasp what the evidence shows. Jaspers is not alone in his observations, which collate a whole series of such. Joseph Needham, in *Science and Civilization in China*, notes:

> The close coincidence in date between the appearance of many of the great ethical and religious leaders has often been remarked upon: Confucius, c. -550; Gautama (Buddhism), c. -560; Zoroaster (if a historical personage), c. -600; Mahavira (Jainism), c. -560, and so on. But the Chhun Chhiu period was also contemporary with many important political events, such as the taking of Nineveh by the Medes in -612, the fall of Babylon to Cyrus in -538, and the invasion of the Punjab by Darius in -512, all examples of Iranian expansion. At the beginning of the Warring States period, the Greeks checked Iranian expansion westwards (-480), and the middle of the -5th century saw the erection of the Athenian Parthenon. The concluding stages of the Warring States time are contemporary with many outstanding events, such as the conquest of Alexander the Great (c. -327), the foundation of the Maurya dynasty in India and the beginning of the reign of Asoka (-300 and -274 respectively), and the Punic Wars in the Mediterranean (-250 to -150) which overlap with the first unification China under Chhin Shih Huang Ti. But the beginning of the Roman Empire (-31) does not take place until well into the Han dynasty.[10]

These observations began earlier in the nineteenth century as global historiography began to force the issue of a multicultural perspective, and this entailing the need for synchronous study. The first philosopher of history to mention the Axial phenomenon would appear to be the little

---

9 From Karl Jaspers, *The Origin and Goal of History* (New Haven: Yale University Press, 1953), Part I, Ch. 1.
10 Joseph Needham, *Science and Civilization in China* (Cambridge: Cambridge University Press, 1965), p. 99.

known Lasaulx (1856), who observes,

> It cannot possibly be an accident that, six hundred years before Christ,

Fig. 2.22 Archaic Greece

> Zarathustra in Persia, Gautama Buddha in India, Confucius in China, the prophets in Israel, King Numa in Rome and the first philosophers—Ionians, Dorians, Eleatics—in Hellas, all made their appearance pretty well simultaneously as reformers of the national religion.

A sense of something defying probability arises spontaneously as we notice this phenomenon. Victor Von Strauss (1870) notes,

> During the centuries when Lao-tse and Confucius were living in China, a strange movement of the spirit passed through all civilized peoples. In Israel Jeremaiah, Habakkuk, Daniel and Ezekiel were prophesying and in a renewed generation (521-516) the second temple was erected in Jerusalem. Among the Greeks Thales was still living, Anaximander, Pythagoras, Heraclitus and Xenophanes appeared and Parmenides

A Riddle Resolved

was born. In Persia an important reformation of Zarathustra's ancient teaching seems to have been carried through, and India produced Sakyamuni, the founder of Buddhism.[11]

We can now return to consider the Greeks, and note that many observations of the type collected by Jaspers exist for isolated instances of what we can see is connected to this 'Axial Age' phenomenon. Thus the philosopher Bertrand Russell opens his *A History of Western Philosophy* with an exclamation of wonder at this generative era:

> In all history, nothing is so surprising or difficult to account for as the sudden rise of civilization in Greece. Much of what makes civilization had already existed in Egypt and Mesopotamia, and spread thence to neighboring countries. But certain elements had been lacking until the Greeks supplied them…What occurred was so astonishing that, until very recent times, men were content to gape and talk mystically about the Greek genius. It is possible, however, to understand the development of Greece in scientific terms, and it is well worthwhile doing so.[12]

We suddenly see the question of Greece in the larger context of the Axial Age, and to understand the question in scientific terms requires an objective look at a phenomenon that we had not suspected, where the occurrence of so many novelties in parallel seems at first inexplicable. In any case we are left with a question, is there a science of history?

The implications of the Axial Age have thus left its study stranded in a kind of limbo, as the phenomenon has tended to drift into misinterpretation. Karl Jaspers, in a curious blend of the religious and the secular, brought a carefully balanced sense of the philosophy of history to his depiction of the question, but many in his wake have tended to see a kind of generalized 'age of revelation' in which the issue of religion is given center stage. And this has tended to scare away serious students of the subject. But if we examine the data of the Axial Age more closely we discover to our surprise that it is more than just an historical garlanding of sages and prophets. If we zoom in more closely we discover to our astonishment that these sages and prophets are merely the tip of an iceberg, that the Axial phenomenon encompasses an entire social transformation in place of an entire stream of culture. And we

---

11 From Karl Jaspers, *The Origin and Goal of History* (New Haven: Yale University Press, 1953), Part I, Chapter I, "The Axial Age".
12 Bertrand Russell, *A History of Western Philosophy* (New York: Simon & Schuster, 1945), p. 3.

soon see that the question of religion is only one aspect of the mystery. For as the remark of Bertrand Russell suggests the case of Greece comes to the fore in the synchronous emergence in parallel of multiple Axial exemplars, and leaves as its clearest case the spectacle of secularism at the point of its

Fig 2.23 Temple Ruins

birth in world history.

As we examine the Axial Age in its breadth we are confronted with the difficult question of arriving at the history behind each of its exemplars. Thus the history of India behind and leading up to the remarkable era from the appearance of the Upanishads to the birth of Buddhism is difficult to reconstruct. And yet the basic outline of the Axial phenomenon is clear. And the question of what is historical in the Old Testament at first bedevils any simple account of the birth of that remarkable document. China, in turn, while it clearly echoes its parallel cousins, confronts us again with a confusing picture of the period in question. Ironically, then, despite the hopes of religionists for some secular version of the idea of an 'age of revelation', the clearest example given to us, the period of the Greek Archaic onward, shows us in detail something quite different, and in many ways far more remarkable: a kind of evolutionary leap or jump to a higher level of civilization, one very well balanced between all the categories of culture.

The notion of the era of Classical Greece as the birth of the secular would at first seem paradoxical. We need not press the point save to note that the birth of philosophy as a critical consciousness sows the seeds of rationalism for the first time. In fact, a balanced view is essential, for the essence of the Greek phenomenon could as well be argued as the last flowering of a strange form of political polytheism, and we should be wary of assigning a modernist label to what we see. But the gestation of philosophical tradition in Greece shows us the first birth of the Enlightenment, as it were, along with

Fig 24 The triple transformation of the stream of ascetic Indian religion to the sequence elements of yoga, Jainism, and Buddhism is a tour de force of the Axial interval.

the first birth of science, the first Scientific Revolution millennia before the one that centers the transformation to the modern world in the sixteenth and seventeenth centuries. The point here is that the Axial phenomenon is clearly connected to a larger set of categories than the merely religious, a point that is clearly indicated in Jaspers' original description, although he is struggling in the text of his work on the subject to remain without his theological boundaries, and yet to see that something larger is at work than the legacy of Christian historicism. Axial Age Greece was a multidimensional masterpiece whose legacy has ultimately transformed world civilization.

As we look beyond the pointillistic sprinkling of great minds in the Axial interval and examine the question of what happened to the culture as a whole we begin to see that there is a kind of transition in a cultural totality leading to a new and more advanced stage of civilization. The Greek phenomenon thus crystallizes as new cultural substrate in its Dark Age, then begins a kind of take-off in the Archaic period beginning in the eighth and ninth centuries. We see a field of city-states emerge in a spectrum of political experiments, as dramas of class struggle and republicanism yield finally to the first great democracy in world history in the case of Athens. Pervading this general tide of sociological rebirth is the manifold of cultural achievements that we associate with the classical era, from the creation of the Homeric epics from an oral tradition, with a great flowering of poetic art climaxing in the birth of the Greek tragic genre. We see the birth of philosophy, and science, and, indeed, the birth of historiography in the works of Herodotus and Thucydides, and others. The entire account of the Greek achievement here is then something far larger than the individuals that make it up and constitutes a kind of eerie time-slice of creative upheaval, one as remarkable in swiftly coming to a close as in the suddenness of its arising.

In fact the dates suggested by Jaspers for his 'Axial Age', -800 to -200 seem overly generous, for we can see, if we take the example of Greece as a defining instance, that the interval of great innovations is essentially over by -400, and that the onset of the Hellenistic period is of a quite different character. This is clear from the way the great experiment in democracy yields to the resurgence of empire in the conquests of Alexander the Great whose legacy is to create a larger oikoumene into which the achievements of Greek civilization diffuse. We thus are confronted with an interval of the Greek Axial Age that almost suggests a kind of 'punctuated equilibrium', to use the phrase of the evolutionists, for we can almost clock the 'punctuation' in the brief period from the late ninth century to the generation of Plato and Aristotle, followed swiftly by the seeming 'equilibrium' period in its wake as history seems to resume its less spectacular course.

> By our reckoning using our model the Archaic Age is the key interval, but we see an immense flowering in its wake, relatively late. Is our model off? The most we can hope for with a crude model is a rough insight: we are not sure here, but...the model clearly points to 600 BCE as the 'divide' line to the transition. And mirabile dictu, just as in modernity, with its democratic flood at the divide, the ancient case shows the sudden appearance of Solon at the divide. We know by 'taste' this isn't a coincidence. Note the way 'system action' correlates with the seeding of emergent democracy, while the post divide interval as 'free action' is the actual realization....

While many who have attempted to grapple with Jaspers' framework of an Axial Age have narrowed their focus to the issue of religion, we can begin to suspect, to the contrary, that the case of Greece suggests something broader. And if we take to heart the case of Archaic Greece, and look at the emergence of Israel, we begin to see an analogous period of social transformation that just so happened to produce the seeds of what was later to become a series of monotheistic religions. It is important to see that the

Fig. 25 Solon the lawgiver

history of Israel in the Axial period at least is that of a Canaanite culture and its passage through an age of empires, as it creates an epic literature of itself, and leaves this in its wake, as a set of seeds that will, as with the case of Greece, diffuse into a larger oikoumene. We can begin to see the structural similarity between the two histories, and to notice what is most surprising, the way in which whole literatures seem to come into existence in a a strange timing, that of the Axial Age itself.

Later we can attempt to grapple with the parallel histories in India and China, but already we seem to have a basic clue: the general stream of historical emergence is punctuated with a set of innovations that pass into the larger field of history to influence a later oikoumene. The effect is obvious in both China and India, where a close look might also resolve the two harsh contrast between the religious and the secular. For the effect as a whole shows clearly the way in which categories are fluid, as philosophy becomes religion, and religion becomes politics, and politics becomes

Stream and Sequence Consider the dynamics of the Greek or Israelite Axial intervals (or any other for that matter). A stream history leads up to the Axial interval and shows transformation. This transformation generates a higher level step in a greater eonic sequence. This is the 'stream and sequence' effect. We now have two levels to our account, the evolution of the stream of cultures, and the evolution of the high level sequence. And this allows us to give expression to ideas of evolutionary directionality and progress at the higher level. Or perhaps progression would be a better word. However, the idea of an eonic sequence allows us to proceed without committing ourselves on generalizations about progress which always end up confronted with various contradictions.

Culture Streams We can think of the historical timeline or streaming of cultures as their continuous chronicle in time, e.g. the Greek stream: the total history of Grecian culture from primordial Indo-European times to the present. The intersection of this stream with the eonic series in the Axial interval produces a distinctive burst of macro-history. We can consider any subset, superset, or other cultural variable in the same way, the science stream, the history of science, the poetry stream, the technostream, technological history, the econostream, the history of economic systems, etc,…

Economic Streams Note that economic history is distinct from the eonic sequence. Economic activity is continuous and globally omnipresent, while the sequence is intermittent. We are coming to see the problem with the 'economic interpretation of history': it is a dependent process. Note that the explosion of the Industrial Revolution occurs when an econostream intersects with the eonic sequence.

The Eonic Sequence Our non-random pattern is clear: we see a macrohistorical sequence associated with the emergence of civilization in a long frequency or directionality, analogous to (although not the same as) feedback, able to act on cultural streams in intervals of several centuries. We can reverse-engineer this data with a question, Does world history show evidence of any kind of sequence? The answer is yes, and we see very strong correlation with an intermittent sequence pattern that can only be called 'evolution'. This sequence is intermittent and intersects with the various streams of culture it finds in its direct path. This sequence can show synchronous parallelism, and follows a frontier effect, as we will see, and works in a kind of leapfrog effect.

A Riddle Resolved

'sacred'. From Confucius to the prophets of Israel, to the philosophers of Greece and India, we sense of continuous spectrum of realization that is in a most spectacular display of historical dynamics producing a new whole new epoch of civilization in its wake, as this takes the form of a series of reborn 'civilizations'.

2.6.1 A Spectrum of Transitions

We are confronted with a synchrony of effects. We have at least five seminal areas suddenly showing characteristic 'pivotal' intervals in concert:

Archaic to Classical Greece The period from the Greek Dark Age to Alexander contains the great clue to world history. The period of Archaic Greece overflowing into the Classical period lays the foundation for a whole new order of civilization, and produces the beginnings of philosophy, science, and democracy.

Histories of Israel The phenomenon of 'Israel', that is, Israel/Judah, in the Old Testament is a considerable enigma but its significance falls into place once we see that it simply reflects its place in the Axial phenomenon. This involves the period from about -900 to the Exile, and does not include the (mostly mythical) accounts of Abraham to Moses. No historical myth, theory of evolution, or universal history has ever produced a coherent account of this history. But the macro effect will clarify its status at once, and in a very simple and elegant way, if we see that the key issue is the core period of the Prophets around which additional history is adjoined as epic prelude.

Persia As we study our data we begin to that the innovative areas are almost always at the fringe or frontier of the main centers. Thus Israel/Judah is a remarkable upsurge in what were then the frontier areas of Egypt and the Mesopotamian mainline. Even so, we can see that the parallel development of monotheism in the legacy of Zoroastrianism suddenly blends with the Israelite during the Exile, producing a monotheistic corpus on the threshold of its global religious formation.

China: The period of Confucius One of the strangest cases of the 'axis' effect is the sudden transformation *in medias res* of the Axial period in China. This comes right on schedule in the midst of an otherwise continuous history! The rise to organized states in Chinese civilization begins very early, and yet we see the synchronous effect right in the correct time frame, as an overlay on the prior development. China and Europe are both at the fringes of the 'macrosequence', at this point (we notice nothing in Europe). The Chinese case is inexplicable in isolation. This shows that the Axial/macro effect occurs on schedule independently of the local dynamics of civilization.

India: Upanishads to Buddhism The case of India resembles that of our 'Israel' in producing a world religion from the temporal sequence, as if sifting from a tradition that is already clearly formulated (relative transform) and existing prior to the transition. We see that some dynamic is operating independently of the politics of cultures and empires in the reactions of religion to state integration. With the forest philosophers who renounce history, India creates a protected zone, a parallel world in the Axial spectrum.

Early Rome We should include the case of Rome either by itself or as a cousin of the Greek case. Note that when we speak of the Greek period we are referring to a network of city-states stretching all the way to southern Italy. The appearance of Republican Rome in the wake of the Axial Age is prime data for the 'axis' effect. Note that the Roman Empire is a much later phenomenon, and in fact dramatizes its own deviation and decline from the sturdy Republican beginnings appearing in the Axial interval.

The New World and Africa Since this phenomenon is global we should wonder about areas where the evidence is absent of an Axial effect. But if we examine the onset of the Maya we see a structural synchrony as a relative transformation in sync with the rest. We can draw no hard conclusions here, save only to note that nothing in the South American data contradicts our portrait, and the few elements we have fall into place. In the case of Africa, we should realize that the continent was until very late in the stage of the San hunter gatherers, and that the expansion of the Neolithic and the Bantu migrations were barely underway, so we should not expect sub-Saharan regions to correlate necessarily. Egypt, of course, is one the great sources of higher civilization to come. It is important to grasp the difficulties of survival in Africa, until the coming of modern medicine, and we should for these reasons see that, while Africa and Eurasia are a continuum for our analysis, the case of the African interior is problematical at the earliest phases of our data. But the case of Africa will make sense once we expand our data to a sequential pattern: then it becomes obvious that the Neolithic phase is the beginning of civilization in Africa, notably with the migrations of the Bantu peoples, who are the first wave of higher civilization to the sub-Sahara. In the case of the New World this issue cannot easily be resolved since we don't really know to what extent diffusion from the Old World has taken place.

A Riddle Resolved

Fig. 26, Homer, Old Testament scene: Moses found, *Mahabharata* scene: Krisna/Arjuna, *The Wrath of Achilles*.

The triple transformation in parallel producing three epic literatures in the Axial interval is a remarkable macro-incident in the Axial Age. In each case semi-historical figures become epic heroes.

In each case we must distinguish the prior stream elements and their Axial transformation in the Axial interval.

2.7 The Axial Transitions: Stream and Sequence

We are beginning to sense that the Axial Age is something more than a cluster of brilliant sages: it is a transformation of a whole cultural sphere. We can also look at it in terms of our ideas of stream and sequence: a series of parallel streams suddenly intersect with a larger sequence. We begin to notice the Axial pattern in terms of creative individuals. But its philosophers and sages are the tip of the iceberg, and behind them we see whole cultural regions, out of the blue, proceed rapidly to a new stage of culture. On the one hand it is essential to induce change through individuals seeding cultures with new ideas. But this does not explain the overall coordination over time of complex emergent events, i.e. a string of poets, the birth of democracy, or a world religion. The clearest and best-documented case is that of Archaic/Classical Greece. Roughly we have the following remarkable surge:

Early tribal history of the Greeks
1200-900   From the Mycenaean period to the Greek Dark Ages
900-600   Dark Ages yield to Archaic period
600-400   The great take-off period, the Greek 'Miracle'
400 onward: We enter the Hellenistic, it's over

Compare this now to Israel:
Early Canaanite 'stream' history
1200-900?   The onset of the 'Israel/Judah' kingdoms
900-600   The history of 'Israel/Judah', emergence of Prophets
600-400   The Exile period, crystallization of texts
400 onward: A new religion has come into existence

Note the remarkable similarity! This periodization is slightly formulaic. The interval from -900 to -400 encloses the basic mystery, the intersection of stream and sequence. But the three centuries from -900 to -600 show the field of action, with a 'divide' point about -600, as the 'output productions' proceed from the 'Axial' causation. Similar periodizations break through their disguises in Rome, China and India. Note the isomorphic character of these two histories. It is impossible to distinguish 'sacred' and 'secular', once we see the connection. This is the embedded 'transition' pattern. In the middle of continuous stream of culture a sudden relative speed up occurs. We must realize the high level at which this dynamic is operating. And remarkably in both cases a great literature comes into existence, the Greek and Old Testament epics. Note that 'Israel/Judah' disappears near

A Riddle Resolved

the Exile, depriving us of a flowering or realization period, but enforcing a mysterious extra-state character to what will be a 'cultural complex' that travels transculturally. Note that our system treats states and religions equally in its dynamics, and it is the case that the core Axial period for 'Israel/Judah' is about a state, and not about a 'world religion'. Judaism as we know it comes much later, as do Christianity and Islam.

The five centuries from -900 to -400 enclose the principal effect. But the real interval is very early, from -800 to -600. This periodization in turn shows us what's going on in the case of Israel. The case of Archaic Greece is especially telling because we aren't distracted by religious questions. Its clarity is enhanced by the fact that its earlier Mycenaean phase collapses and Greece goes into what is called its Dark Age. This is the 'stream and sequence' effect. The stream of Greek cultures shows many civilizations, but that part that we call the 'Axial interval' stands out. Why? Because, as we shall see it intersects with a greater sequence. In the same way the Old Testament is confusing, because it includes, just as does the *Iliad*, the prior tales and chronicles of a Canaanite people, the stream history, but the crucial era is the Axial interval and this occurs in exact concert with that of Archaic Greece. It is clear that the Israelites were aware of something strange happening in their history: they noticed their 'Axial' transformation.

Fig. 27 *Moses Breaks the Tables of the Law*, Dore

The sudden reawakening of Greek history in the Archaic period after the collapse of the Mycenaean is the classic clue. Thus its sudden resurgence, as if on cue and in parallel with the other regions of the Eurasian continent, is the more remarkable. We are stunned, by zooming out to see the whole pattern, to see that the Greek Archaic is exploding *on schedule in a larger system*, not solely as the result of antecedent influences. Against the backdrop of world history as a whole this brief period from around -900 to -400 induces an immensity of innovative advances, with an intensity that has never been matched to this day, although the rise of the modern is a fair competitor. *Note that we cannot ascribe simple causal/local influences as the cause of this phenomenon, since we are observing a part of a total Eurasian phenomenon.* Each of five regions, Rome, Greece, Israel/Judah, India, and China, has an analogous interval in this fashion (the Roman being a bit late), although the Chinese and Indian are less well-known. The case of (early) Rome is really an aspect or variant of the Greek case, and can be considered, at this level of generality, in the same category. We can see that Rome arrives a bit later, for the obvious reason that it waits on diffusion from the Greek system.

We have the clue to the Old Testament: it contains a core account of precisely this interval, with a great deal of other material tacked on as lead-up history. The Axial interval is more about the emergence of the Bible than of Israelite history. Much of the content of the Bible distracts us from the crucial Axial interval. The tales of Abraham, Moses, the Exodus, are clearly the mythical lore of a Canaanite people who, especially during the Axial interval, accumulate a literature that, at the end of this interval, becomes what we know of as the Bible. Note the resemblance of this to the way the Greek *Iliad* and *Odyssey* come into existence. Achilles and Odysseus are not Axial figures, but the crystallization of the Greek epics is a prime Axial effect! The Homeric period in the eighth century is followed by an immense flowering of literature, which becomes incandescent in the period of Greek Tragedy. The latter lasts barely a century, and is gone. Stand back from this analogous biblical phenomenon, which casts its spell to this day. Is it not a very odd and historically embedded text? We see that its correlation with the Axial interval reveals at once its real significance. Our approach almost does better justice to this history. It makes almost no sense stripped of its religious mythology until we see its Axial context.

Note the way the historical facticity behind the Biblical myths changes its character after around -900 in the wake of the David/Solomon era myths and we have the histories of Israel/Judah or general Canaan up to the Exile. As foundational religious history, this is quite peculiar stuff, but in light of the Axial context it becomes precious historical lore indeed, the first documentation in writing of the genesis of a religious formation, a world historical moment. Note how various traditions of prophets suddenly without warning turn into the remarkable string of the classic Prophetic cluster in concert with the Greek timing. The sameness in difference from the Greek example is beguiling. We note how *just around the Exile* the Biblical corpus comes into something resembling its final form, and that from -400 onward the phenomenon is essentially finished, as the record is codified. The seeds of a 'world religion' or the materials for several have appeared and crystallized. Remarkably, as it happens, 'Israel/Judah' suddenly disappears at the period of the Exile, and we don't see the analogous sudden flowering that we find in Greece after -600. It makes no difference, however, to the overall effect.

Fig. 2.28 Homer reciting *Iliad*

There is an irony to this circumstance, it almost feeds the phenomenon itself by separating a literature from a region!

Let us not forget that our discussions of religion in the abstract can forget the obvious: the Bible records the actual history of a Canaanite kingdom during the Axial interval. The religions arise later. Once that interval closes, the record stops. From that point onward, a tradition is born and its adherents are looking backward. The material, here as in the other cases, flows outward into its environment to have what effects it will have. For the peoples of that era, the Biblical corpus was a tremendous new cultural asset, an almanac of Civilization. It spread through the regions of the Roman Empire and beyond because it was of great help in the assimilation of tribal peoples to the shock of expanding civilization. We constantly think of religion in terms of metaphysical abstractions, but that misses the historical point that, sourcing in the Axial interval, a cultural instrument appears that assists the process of tribal integration into the world system. For its time the Bible was 'state of the art'.

We can be certain that the Biblical history is only one aspect of what must have been a far more complex 'Axial interval' in the regions of the Middle East as a whole. As our strange phenomenon shows it is trying to balance itself, as it were, across Eurasia. In the context of monotheism, the phenomenon of Zoroastrianism, and much else, should join our account. It

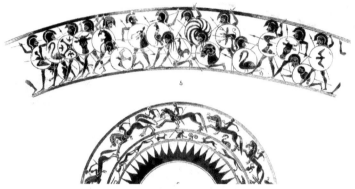

Fig. 2.29 Perfume bottle, Greek Archaic

is not clear just when Zarathustra lived, but it is highly significant that just at the Exile there is a blending of the Biblical and Zoroastrian literatures. Be wary of thinking that this period *invents* monotheism. That probably already existed, and is in fact a primordial belief fairly well known to the Paleolithic in the Great Spirit cultures, for example. What we are seeing here is the effect of the Axial interval on what is, as primordial monotheism, already probably in existence, mixed no doubt in a melee of polytheistic beliefs. That effect is to spawn a world religion. But the Axial Age as such has nothing to do with religion, and in fact the term requires adjustment to historical realities: we see that religion *in one sense*, and speaking in broad strokes, is simply the ad hoc output of the Axial interval. It is important to see that the actual world religions we now speak of *are constructs outside of the Axial Age*, created by men recalling unsuccessfully the histories of that period.

In general the whole case of the Axial interval in the Middle East must be larger than what we see now. But the case of Israel/Judah is put in writing, and somehow carries the day because, and this is the whole point, it produces a literature and cultural record that outlasts everything else from this time and zone. We will brutally secularize this history, thus bringing out its true beauty, but let us note forget how remarkable Israelite history is. Just as the

A Riddle Resolved

Israelites suspected, there is a Big History at work. And we have stumbled on something else of significance. As we move to complete our pattern, we will see that Sumer and Egypt comprise an earlier version of this kind of phenomenon, millennia before. But these centers of earlier advance are silent in the Axial Age. Why is that?

The clue here is to see that Greece, and 'Israel/Judah' are essentially frontier areas, for their time, and that the drama of the great Empires, such as the Assyrian, which are the legacy in decline of these earlier periods, is being bypassed by clusters of innovation on their fringes. These empires are dinosaurs of an earlier age period. Note how the Greeks barely survived these attempts by Empire to destroy their world, that the 'Israel/Judah' is a dramatic account of the history of surviving or not surviving these destructive empires. The Axial Age here is a record of innovation outsmarting the momentum of the legacy of a previous cycle of civilizations. This *frontier effect* will help resolve one of the puzzles of the isolation of the rise of the modern on the fringes of the European system. Thus if you factor out the Sumerian core area, and the Egyptian zone, and consider the Occidental zone of the Eastern Mediterranean and South Asia, you suddenly realize there are very few innovation zones available. One is precisely the Canaanite region, another the Greek, conveniently buffered by Aegean. The Indian, Chinese, and Roman cases automatically fulfill this requirement also.

We have the essential framework for what is happening in the Roman, Indian, and Chinese cases. In many ways the Roman phenomenon is part of the Greek, which spawned an immense network of city-states and republican experiments, from the Black Sea to southern Italy, some of them producing democracy, the classic Athenian. Thus Rome springs into existence in the wake of this network and is essentially a variant of it. We must keep in mind the obvious fact that all these cultural zones are extremely different in character and that we can't be talking about the autonomous mechanics of these cultures or civilizations. The Axial phenomenon simply happens independently of the prior histories and cultural mechanics of each region. In the Greek case we do see, however, the expenditure of Axial impetus on innovative cultural forms, among them Greek democracy. Note that this appears suddenly in the wake of Solon, -600, and that it doesn't last very long. Later we will realize even these particulars, down to the decades of a half-century interval, are not accidental.

2.7.1 Archaic Greece: The Clue

Our stream and sequence metaphor is especially apt, and illuminating, in the case of Greece, which has both a long stream history, and an intersecting history in the Axial period. The whole effect is almost eerie and, furthermore, shows us the real key to parallel history of Israel/Judah, strange as that might at first seem. The Greeks would seem to have separated from their Indo-European ancestors in the period ca. -2000, and then entered Greece to stage the Mycenaean civilization.

Fig. 2.30 Relief of tragic mask

1800 to 1400   Cretan and Mycenaean civilizations

1260 to 1230   Mycenaean attack on Troy VIIa

1200 to 1050   Dorian invasions, a Dark Age begins

From 900   Axial Interval to about 400

900 to 750   Emergence of *polis*, the spectrum of Greek city states

800 to 700   Greek alphabet and the work of Homer

650's onward   The first 'age of revolution', republican *poleis*, Solon,…

500's onward   Late emergence of Athenian flowering, democracy, tragedy, a scientific revolution, philosophy, and much more, cascade in a spectacular display

400's onward   Clear waning of transitional effects, coming of Empire phase

The discovery of the Axial Age by Karl Jaspers and others was one of the most important achievements of modern historiography, but the result has often been a series of misinterpretations of this phenomenon, and an inability to escape the framework of Old Testament history.

The terminology of the Axial Age has devolved into a confused perception of some kind of religious age, a sort of generalized age of revelation. Indeed! But not in the sense intended. And this Old Testament fixation has resulted in the inability to see the phenomenon for what it is. The phenomenon of Axial Age Greece is then seen as in some fashion not conforming to the

A Riddle Resolved

World Line of The Eonic Observer

Short of a science of history, we need someone to be a simple observer of the macro or eonic effect. We can call him an 'eonic observer'. He can definitely aspire to a science of history, and to be a Universal Observer, and yet that is the whole point, he is limited by the time and circumstance of historical immersion, and ideological participation. The first and most telling example of the eonic observer lies with the redactors of the Old Testament, who were unwittingly observing the Axial Age. Our observer can be immersed in history and still record 'eonic data'. He should graduate to ever larger data sets and be collecting data over many millennia, at the end of which he starts to do theory. It would be nice to be outside of time, or in a rocket module in orbit, going into suspended animation during off periods. In fact, he is embedded in history, and going through paradigm changes in each of our transitions, executing scripts in each revolution. That is all of us. Every time we use the term 'modern' we are observing the eonic effect, looking backward. We are all eonic observers. We use terms like 'rise of the modern', the 'middle ages', the 'age of revelation', and so on. For real science we should be objective observers, assessing data to be put in a time capsule until the end of the eonic sequence, if ever. The last eonic observers, if any, might have a hard time seeing how the data was filtered through the local paradigm of his previous incarnations.

Thus, we can make a formal idea out of the observer of the eonic effect. We can invoke the image of an 'eonic observer', with a serious or humorous image of a scientific type, jungle hat, library card, lab smock and clipboard, stop watch, rocket ship, anthropologist and time and motion man of civilization, with his atomic stopwatch designed for time measurements on the order of millennia. One more piece of equipment: a paper stamp labeled 'eonic Data'. Wishing to be a neutral observer, he finds himself temporally bound, and his theories prone to become scripts to create further history. We need with some urgency to apply that paper stamp to the Old Testament, 'eonic Data'. The point is also that the observer and his observed history cannot be separated in any attempt at a science of history. WHEE

archetype of an age of revelation, and ends up the black sheep of the Axial Age. The reality is that the study of the Greek Archaic is the key to seeing the real Axial effect, undistracted by questions of the emergence of religion. And the irony is that by studying the example of Greece we can find the clue to understanding the highly confusing history of Old Testament Israel. The interval of Axial Greece is one of the most enigmatic of historical periods in the way it suddenly spawns a fast run of creative innovation, and this, as we zoom out to see the context, almost like clockwork.

The Biblical history has been so overdramatized by epic supernaturalism that we can no longer see what the history was, or its significance. If we turn to Greece it is like catching something unexpected in the act, and in the end far more remarkable than the embroidered sagas of the Bible, now seen in many cases to lack an historical basis. Simple periodization and a bird's eye view of world history as a whole gives us the indication of something very strange: if we track changes in centuries relative to millennia, the whole history of the Greek phenomenon looks almost miraculous, as we note the overall pattern. Something doesn't add up in the usual analysis. We have the canonical instance of an 'eonic transition'. And in this case we the phenomenon in its full detail.

The unexpected suddenness of the Greek transition is remarkable. In *The Origins of Greek Civilization*, a study of Archaic Greece, C. G. Starr describes the inexplicable and truly extraordinary period of the Greek Archaic and is driven to feel that

> the common historical view on this matter [of the tempo of historical change] is faulty. It is time we gave over interpreting human development as a slow evolution of Darwinian type; great changes often occur in veritable jumps.[13]

As Starr, in a further book on this period, notes at the beginning of *The Economic and Social Growth of Early Greece: 800-500 B.C.*, the Greeks in -800 lived in small rural villages on the Aegean, "three hundred years later Greek life was framed in a complex economic structure embracing much of the Mediterranean and centered in cities which were socially differentiated", creating the foundation of the great classical period.[14]

---

13 C. G. Starr, *The Origins of Greek Civilization* (New York: Norton, 1981), p. viii.
14 C. G. Starr, *The Economic and Social Growth of Early Greece: 800-500 B.C.* (New York: Oxford, 1977), p. 3. Starr also notes the same effect in the first phase of our sequence: in *A*

A Riddle Resolved

There is no simple answer to the complexities of what we are seeing until we start to consider what the broad sequence of our turning points suggests, relative beginnings, and a reworking of the incoming stream. This means that, while many genuine novelties are appearing, by and large, we see a transformation of what is entering a period and what is emerging. The dynamic seems independent of the content. Things appear in a total cultural

Fig. 2.31 Noah's Ark

spectrum, with Greek philosophy and early science, dramatic tragedy, or pottery, showing the passage from one end of the spiritual to the other of art, politics, and economy. The key is that the interrupt is coming on cue, and simply creates a kind of intensity or amplitude of generative change.

We are forced at once to distinguish two different things:

> the temporal ongoingness of cultural evolution, a 'this leads to that' aspect,

> an interrupt phase: fast action, accelerating from earlier periods.

History of the Ancient World, he traces the steady development from the Ubaid and Uruk and describes the sudden change in the period just before -3000 by noting that in history there are "revolutions as well as slow eons of evolution; one of the greatest explosions now took place and affected virtually all phases of life in an amazing, interconnected forward surge."

Consider Greek history in this light. We have a people, its temporal sequence, a series of stages, nomads arriving from Asia, early Neolithic farmers, Bronze Age Mycenaeans, then suddenly the period of Archaic Greece, and its Classical ascent-vertical as a foundational period that templates a whole new age. We see this five times, at all once, to the century, in some cases to the decade. The sudden advance of the Greeks does not spring, then, from long antecedent influences, although the raw material of diffusion is there. This means that it happens suddenly without slow buildup, relative to the scale of intermediate mideonic stages, even as it must accept the antecedent influences of a long runway, whose only effect can be timbre but not the note.

The Greek example, especially, shows the spectacular surge, then its first flowering, roughly, after -600, as science, drama, architecture and sculpture, political thought, and a Mediterranean presence, and much else, emerge,

Fig. 2.32 Yogis under banyan tree
Tavernier's *Collections of Travels Through Turky*
into Persia and the East-Indies (1688)

develop, and create whole new categories of thought, social existence, and art. We can break the problem down into clear stages, relative to world history, stripped to a minimum of actual data.

From -900 onward, there are barely visible signs of Greek renewal as it appears from its Dark Age. There is a pronounced appearance of a new pottery style, the Geometric. By the turn of the eighth century, the onset of the earliest period of what is called Archaic Greece. The record of the Olympic Games begins in -776. By the end of the century, the take-off is gathering momentum. Out of nowhere we find the *Iliad* fully accomplished as a written epic, Hesiod following in its wake, then a great flowering of poetic forms. The Greek city-states are crystallizing in an era of colonization, social revolution, and economic advance. By the middle of the seventh century, a new form of culture has arisen, one in which the early Sparta, and Athens, are still cut from the same cloth, a generalized field of city-state constitutionalism, with a trend toward republicanism. At the rough era of the Exile, we find, in the generation of Solon, ca. -600, the Archaic Age graduating, the labels are relatively arbitrary, to what we call the Classical Period, the age of Marathon, Herodotus, the birth of Greek Democracy, Pericles, and the Parthenon, and the Peloponesian War. Soon, by the fourth century, we are in the age of Plato, Aristotle, then Alexander, and the rushing advance wanes.

We see this basic structure repeated in each case, China, India, the core Old Testament period, and Greece. Persia, indeed Assyria, Rome, and other areas such as Carthage, perhaps, are slightly different, but clearly related, variants. The cultures in the original core area, like Assyria, tend to *fail* because they are too large, retrograde or caught up in the past. It is the nimbler Israel and Greece that take off. Analysis requires great caution: the overall perception of a *mechanical* event is rendered over to correlation by a seemingly random pattern of *creative* events. It seems like a 'spiritual' phenomenon. Confucius, Laotse, Buddha, Mahavir, Deutero-Isaiah.

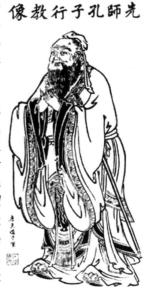

Fig. 33 Confucius
Cf. *Confucianism as a World Religion*, Anna Sun, Princeton, 3013

The Hellenic example is of especial interest because its stream shows so clearly the four or more separate conditions of culture possible to the nomadic tribalisms entering the field of successive phases, in the relations of multiple encounters with the eonic sequence:

1. its earliest stage as a nomadic tribalism arriving from Asia and Hyperborean minus infinity. By what process of cultural evolution the early Indo-Europeans achieve their characteristic culture remains unknown. The same stands true for all of the primordial cultures of the Paleolithic.

2. Then, a sequential or mideonic stage in the first phase of civilization after Sumer, as the Mycenaean relative and apprentice of the Minoans. The difference between a *phasing transition* and the *sequential dependency* induced it its wake is clear from looking at the Mycenaean world, very much in the mold of the Middle East, and the Minoans, themselves in a complex blend of this same, and earlier diffusion. This era makes what comes later the more remarkable. For it shows that pure diffusion is a different effect.

3. a phase of eonic transition: after an artificially created or contingent 'Dark Ages', we see the rapid appearance of the transitional period leading to its great classical contribution, followed by

4. a post-transitional passage into its Hellenistic period as a generator of a new oikoumene.

This is not the evolution of a 'Greek' culture, but eonic evolution in the greater eonic sequence, in a cross-section or cycle sampling, during a period of phasing transformation. This is confusing because a process universal in scope exploits the tribal/local to refresh itself and create new templates of cultural advance that will then find themselves short in the passage to their real destiny, the molding of oikoumene cultures, that don't have this phase intensification, into an integrated whole. It is hard to avoid the conclusion that a *local* acceleration finds its meaning in a *global* context. The sudden transformation occurs just as the great cycle of phase picks up, and does so in a 'near-far' relation to the nearby Mesopotamian world. This 'near-far' is the mechanics of parallel interactive diffusion. The transition induces *more* interaction from a safe distance, during the Orientalizing period in the seventh century.

The case of Greece is especially interesting because of the artificial discontinuity created by its post-Mycenaean collapse. We might be hard-pressed to uncover the identical pattern in China, visible from ca. -750 to ca. -400, without the Greek example. The Chinese example shows that prior growth, relatively strong in this case, is an independent process, a fact that might elucidate the modern period. For any earlier developmental continuity is merely summed with the interrupt phase, which is only visible from its highest achievements. Indeed, Greece is nearly reduced to the Stone Age after the collapse of the Mycenaean period, starts from behind and then overtakes its greater environment! We might try to extend the buildup to -1200 in some particulars, but the very nature of the evidence cautions that an effect is visible only because nature could not manage five separate generations unless its synchronous action were brief, indeed synchronous. The whole effect of this parallelism is extraordinary and yet it has gone virtually unnoticed, or ignored, except among a small string of scholars, and, indeed, has been the object of dismissal by others.

With this simpler Greek example, we can decipher the Old Testament data, without being distracted by religious trappings. It is remarkable how the Old Testament, with an additional account given by later history to the period just after the Exile, gives direct clocking testimony of one time-zone slice, the Canaanite pocket world, to the whole phenomenon of the great synchrony, irregardless of its content. The Old Testament is a series of 'story slots' built around the eonic effect in its core period in the interstices of Mesopotamia-Egypt that its redactors 'knew' without knowing must correspond to their historical record, whose exact details they were hard pressed to reduce to fact. The runway, acceleration, crossing, and realization-emergence are told in the thoughts and words of a crystallizing first-emergent group, the Israelites becoming the Jews in the later Hellenistic world of the Second Temple. In India, the chronological record is not so detailed but as clear, the appearance of early Buddhism in the period after -600, within the memory of the earlier Upanishadic era just before it, is almost directly parallel, bulls eye fashion, within the limits of a generation. Just as the Old Testament literatures begin to crystallize by -400, so the 'Buddhism' we see has crystallized from the fertile era of gestation, in the period before roughly -600. The 'peculiar' appearance of the Upanishadic phenomenon as a buffer between the runway and emergence periods is a giveaway, as incomprehensible as the rest, but the bearer of a clue in the form of its preoccupation with self-consciousness.

2.7.2 *The Old Testament as Eonic Data/ The Persian Case*

One of the most remarkable cases of the eonic effect is reflected in the Old Testament. Historians are beginning to close in on the Old Testament period, to produce an account that finally begins to make sense of the confusing history and scholarship here. Biblical scholarship, so-called, has often been little more than the theologian's disinformation. We have to manage to be somewhat ruthless, and yet respectful here. We are about to annex the Old Testament to a secular model. The document, as it stands now, is beyond salvage on its own terms. But a secular account can fail as badly as the religious.

One advantage of our eonic approach is that we can partition world history into a series of meaningful blocks, and assess their high level relationships, up to a point, without the exact data. Thus we might inject some bogus data from the Old Testament account, passed like bad money by theologians, and then find that wrong. But our 'eonic history' of the Old Testament would remain, more or less. That's because it is pure architecture with default content, e.g. the well-attested facts we know, and even those we may not know. And those facts are almost entirely in the 'eonic Axial range'. Almost nothing can be taken at face value in this labyrinth of distortions. But an invariant structure remains in all accounts. That high-level model merely says that the *core* Old Testament block, a few centuries before the Exile, roughly, in the period of the Prophets, shows 'eonic determination', Axial Age correlation, same as Archaic Greece, which it resembles very closely (at this level of abstraction). We can see immediately on the grounds of periodization alone that we are missing something in the standard accounts, religious or secular. The religious account is mythic, while the secular can't explain the timing. Timing of what? However, the right data finally seems to be emerging, and it fits our eonic model to a tee.

In fact the whole document falls into our lap as a play of 'eonic data' built around a transition, albeit in disguise. Don't be distracted by monotheism here. Like Orpheus, if you look backward at Eurydice, you will be lost, confused all over again. A transition is a fuzzy time-zone patch where eonic emergents appear on schedule in a frontier effect. The relative transform of the nth god name sequence is itself an eonic emergent, monotheism is an eonic emergent self-referentially applied to its own 'history'. A close look shows an embedded account of this eonic transition. Let us look again at our stream analysis of the Greeks:

> An independent stream, e.g. Indo-European Greeks
>
> A mideonic entry into a diffusion field, e.g. Mycenaeans
>
> A transitional time-slice, e.g. the Archaic Greek period
>
> A post-transitional oikoumene

Let us note in passing that the third, transitional period produces a great literature in the gesture of putting the *Iliad* into writing, sometime in the eighth century or early seventh. This literature is about the second Mycenaean period, which is not a part of the Axial period. So it is the

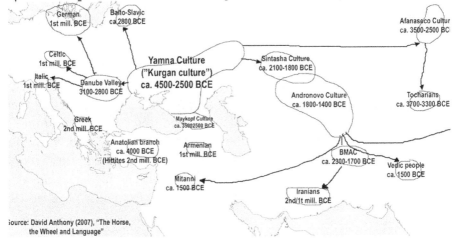

Fig. 34 Indo-European Migrations

transitional rendition of 'stream entry myths' that is significant.

Now substitute the relevant data from the Canaanite area of the emergent 'Israel'. Our Axial period clearly seems to straddle a broad band all the way across Eurasia, one transition in a suitable roughly spaced spot from Rome to China. We have to be careful and not exclude other 'eonic data' in the Mesopotamian region. But, as history shows, this field tends to fail the test of the 'acorn effect' and we see the hopeless cases like the Assyrian empire rise and disappear, unable to extricate themselves from the mideonic empire trap. (Note that Israel is itself barely able to manage its acorn effect, and yet seems to survive its own demise as a kingdom. First 'Israel' is lost, as the remnant Judah becomes the carrier, then that is lost). The only real survivor of this area will prove to be the Biblical documents and the Judaic stream. With that caveat (we will see clear blending later with Zoroastrian thematics), we can take this one great gift of data slightly to the fore. We get the following:

> An independent stream, e.g. Semitic Canaanites

> A mideonic entry into a diffusion field, e.g. tales of Egypt, a kingdom in the field of late Mesopotamian mideonic empires

A transitional time-slice, e.g. 'Israel' and Judah up to the Exile

A post-transitional oikoumene or generator, here spectacular, several religions

The two structures are isomorphic, if we can sort out the actual data that we are dealing with. The Old Testament clearly records a transition, but throws us off the scent because of its instant mythological wrapper. But given this resemblance of our two lists we can safely predict the key period will correspond to the Greek Dark Ages and Archaic period. And that there might be a clustering near the divide, if we can find one to correspond to the modern. Tracking backward 2400 years gives us about -600, the period of or just before the Exile. The clue might lie there and our butterfly net coordinates suggests something interesting between about -900 and -600, especially the last half: about the time of the major Prophets! We check the

Fig. 35 The myth of Israel: Joshua crosses the Jordan

divide period. Let's look at 'state of the art' Biblical Criticism, attempting to uncover the archaeology of Israel. As the authors of *The Bible Unearthed* note,

> During a few extraordinary decades of spiritual ferment and political agitation toward the end of the seventh century BCE, an unlikely coalition of Judahite court officials, scribes, priests, peasants, and prophets came together to create a new movement. At its core was a

A Riddle Resolved

The Indo European Migrations

The next era of the coming Axial Age will show what is tantamount to a whole new cast of peoples, with a remarkable series of diverse transitional cultures, from Rome to China, among them the new cultures of the Indo-European diaspora that will enter the outstanding oikoumene from central Asia. A great deal of nonsense has arisen over the Indo-European question, and a false mystique has arisen as a result. But a closer look shows merely the non-paradox of highly intelligent streams of exterior tribalisms entering the 'sequence' of the eonic effect. The process transforms the entry material and makes it contribute toward the larger history of civilization. The perfect example of that is the stream and sequence aspects of the Homeric corpus which enters the eonic sequence in its transformed glory with perfect timing. This stream and sequence analysis accounts much better for the facts (despite its inherent mysteriousness), once we grasp how to apply the distinction.

The exact details and history of the Indo-Europeans are very controversial and subject to a great many rival scholarly hypotheses, but the basic outline is clear. From somewhere in central Eurasia, probably the steppes between the southern Urals, the northern Caucasus and the Black Sea, the proto-Indo-Europeans, with their characteristic language and culture, began a series of migrations that produced the Italic, Greek, Hittite, Iranian, and Indic branches, among others, that will set the stage for a whole series of new civilizations and their literatures. Their association with the horse and then the technology of the chariot is decisive in their success in entering, and then often dominating, the older sphere of civilization.

sacred scripture of unparalleled literary and spiritual genius. It was an epic saga woven together from an astonishingly rich collection of historical writings, memories, legends, folk tales, anecdotes, royal propaganda prophecy, and ancient poetry.[15]

Fig. 36 Ajanta Caves

So the Old Testament is really a creation of the divide period! It may not be quite that simple, but the point is clear. This is a climax of strains emerging in the period of Axial phasing. Thus the new world of Biblical archaeology is producing a remarkable result, in the almost complete erosion of the standard Old Testament mythology. The secular student of the eonic effect finds the 'eonic rubric', compression near the seventh century, splendidly confirmed by the emerging picture of the rapid crystallization of a viable but still contradictory monotheism in the 'YHWH alone' movement and the testimony of the Prophets, in a rapid phase visible consolidated in the period of Josiah. It is here that many of the outstanding Judaic myths suddenly crystallize via the formation of an ideology of what is still a 'state religion' in the kingdom of Judah. And it is this corpus, complete with its contradictions and the strategies of its lost moment, that will be injected into the world stream, among other characteristics its unwitting record of the eonic effect being the most ironic, and the strange 'miracle' of another kind,

---

15 Israel Finkelstein and Neil Silberman, *The Bible Unearthed*, (New York: The Free Press, 2001).

the secular student must reckon with as he inherits the elegant remnant of this 'tavern of ruin' as eonic data. We tend to get into a snafu over the clear nationalistic origin of the Bible, its Prophetic anticipations (with retroactive fudging), and the final result, which is several religions in tandem. But in fact the whole structural dynamic is 'eonic' from beginning to end, as long as we don't get sidetracked by later revisionism. It is hard to think of anything more remarkable than the appearance of the Prophets, but it is not more remarkable than the appearance of the Greek Pre-Socratics, Buddha, Confucius, and Lao Tse.

We see the pieces falling into place once we realize that the patriarchal myths of Abraham, the tale of the Exodus, the saga of Joshua and the invasion of Canaan, and the Davidic/Solomonic Kingdom are later nationalistic myths emerging over the transition and starting to crystallize just before the Exile. These are stream entry materials from the mideonic period. Elements clearly predated this codification, but the point is that we see the eonic timing almost eerily in place. Who were the Israelites then? In fact we see that current archaeology shows us the highland peoples drifting in and out of Bedouin stages in the millennium before the pastoralist David, around whom a considerable myth is to be created. The account that we have is backdated with the later codifications we now see in the Bible. Monotheism appears relatively late, in organized form, although there is no objection to evidence that it existed in some primordial version much earlier. But there are still clear elements of polytheistic religion until near the end. And in fact, the whole point was that there was a process of consolidation based on the Jerusalem temple, appearing near the end of the eighth century in our 'acorn field', the remarkable Judah.

Now compare this to the Greek case. We can almost map isomorphic elements one to one between the two. Both produce a nationalistic literature during a transition, using elements outstanding from a mideonic legacy of the culture stream. This history of the Israelites turning into Jews shows a remarkable culture-form, something like networking ironically enforced by the repeated loss of the 'geographical base'. The spread of this network into the coming worlds of recurrent empire will prove a source of general innovations throughout that greater area yielding finally to the Roman world, and this feature goes a long way toward accounting for the emergent Christianity to come.

We must be very careful of teleological questions here, keeping in mind that while our large-scale model shows 'eonic directionality', that does not allow us to transfer that directionality to the interiors and their mideonic productions, e.g. Christianity. Our model only allows 'seeds sown in a transition' to create a cone of diffusion in its follow-up, as the period of eonic determination passes into 'free action'. Some other form of explanation is needed. We can make no teleological statements about the relationship of emergent monotheism and later Judaism, Christianity or Islam, save that they are in the oikoumenes generated by the transition. However, we can see that while our eonic effect is intermittent, and complete by the time of the divide, ca. the period of the Exile, the clear sense of the transition is the creation of instruments of cultural integration, oikoumenes, and that is the result we see emerging in the wake of this transition. Beware of teleological thinking here, and indeed we see in the centuries to come clear 'teleological tragedy' in action as the collision and jackknifing of the mideonic and transitional productions. It is worth proceeding to the Indic example to see the eerie isomorphism once again in the transitional gestation and crystallization of a world religion. For a system modeler this result is far more gripping than the mythology of the text itself.

> **The Bible and the *Iliad*** In conclusion, in spite of the dangers of speculation, let us not underestimate our system or forget the implications of our eonic sequence. We just learned to see how remarkable the case of the Greek transition is. It ends up being less equipped to travel culturally than the Judaic, but the core dynamic is the same, and we suddenly are stunned to see a 'frequency phenomenon' behind the rapid emergentism of literatures in the mainline. Thus, as a matter of frequency the *Iliad* appears in world history. What could such a bizarre statement mean? We could backtrack to that period, sure to discover that while Homer might have been a great poet (if he existed at all) historical homogeneity could not be violated, and we could (sort of) imagine how the *Iliad* came about. And yet as we zoom out we see a clear macroevolutionary meaning *in our sense*. Our model can accept this data then, but it is remarkable indeed.
>
> And that does not preempt any other deeper explanation of the context and free activity of a Homer (who might have been a committee). Our eonic periods are truly enigmas. Consider the onset of the Greek Archaic, and the sudden, out of the blue crystallization of its stream entry literature (bards and their oral epics) across the boundary of eonic sequence. Presto, a great masterwork. Thus we can muse on a classic example of an eonic effect, the appearance of the *Iliad*.

This is a frequency phenomenon, no? Regardless of whether we decide on a real Homer or not. Understand this example, and the eonic effect

Fig. 2.37 Rock of Behistum, with inscriptions from *Zend-Avesta*

is yours. The stream, i.e. proto-Hellenic bardic traditions (mixed with other Middle Eastern traditions), suddenly produces a great literature in the wake of Homer, as if on schedule, as it intersects with the cyclical sequence, why? A man wrote this. But it is a clear function of time, taken in our large blocks. So what's the answer? Whatever the answer, we see that the temporal stream and the evolutionary sequence are distinct. What a beautiful way to evolve a field of disparate (and very stubborn) 'primitives', if we can manage the 'nameless something' that does this sort of thing without naming it. Now translate this argument to the Old Testament, and see what you see.

Canaan and 'Israel/Judah': The Old Testament Riddle It is hard, in fact, impossible, to think of *any* other explanation than that of the eonic effect, for what is bequeathed to us by the redactors of the Old Testament, who, incidentally, lived after the events they purported to describe. It is the eonic 'smoking gun', for behind its history, however we reconstruct historical incidents from its account, lies an implicit straddling of the period -900 to -600, with a particular intensity in the period between -750 and afterward, an eonic Bull's eye, and indirect evidence that stands on its own irregardless of the complete facts.

The study of Israel from the eonic perspective is in the final analysis the most effective for it can help in seeing that the impulse to find transcendental explanations is automatically suggested by the intangibility of the eonic sequence.

Minimum Eonic Periodization of 'Histories of Israel':

1. stream approach

2. transitional period: eonic sequence intersection

3. divide period

4. realization period.

That's it, our eonic history of Israel. And it resolves all the paradoxes of the Israel phenomenon. Reflect on the overall dynamic context. The only safe data, as the Greek example might have forewarned us, is that of the prophetic period, precisely at the climax of phase, and the period of the Exile and the post-Exilic history. David and Solomon are almost like Achilles, and Agamemnon, probably existed... The eonic matrix shows us the master key, satisfied by all accounts. The Old Testament redactors in the period from after the Exile unconsciously followed a procedure based on these steps, for the same reason the modern historian is confused by the continuity-discontinuity paradox of the modern, its medieval antecedent, and the sudden clustering near a divide.

2.38 Shiva and the primordial tradition

Thus it is important to see that the redactors were at step 4, overwhelmed by the period at step 3, and attempting to interpret, create, and include the remnant documents and memories of steps 2, and the mythical or semi-historical step 1.

> 1. First we have the 'primordial' semi-historical Abraham/Moses stage, corresponding to the mideonic phase of the Canaanite cultures in the shadow of the Middle Eastern empires springing from Babylon and Egypt, the world of the Ugarit.

> 2. This period of the stream leads into the just-before period of Solomon, the history and kingdom of a people in a not especially extraordinary Mediterranean kingdom and empire, flourishing and then going

into what many describe as a start of political decline. The kingdom is evidently not the transitional phenomenon. By -750 the age of the prophets is the one clear outer symptom of the transition given to us, so parallel with the Upanishadic Age. It is this phase of the prophets that *tokens* the period of transition as such, just as the Greek philosophers token the Greek transition.

3. We see the climax of the prophetic movement just as the divide point is reached. It is indeed extraordinary to see the emergence of monotheism and its sudden packaging in the period after -600. As the system crosses the divide we see the Persian phenomenon and its state 'Zoroastrianism', blending in, and then the great expansion of the Jewish network into the Middle East and Mediterranean worlds.

The ship has set sail, and we are in the emerging world of Judaism. Shot out of a cannon, the Israelites become Jews and burrow into the Roman Empire as a parallel counterpoint to the 'great Athens' passing into Rome. Like a 'throw and catch' in a computer program it is this strain in the great classical phase that will unveil from its latency the 'failsafe' response to the great passage from transitional 'eonic determination' to 'free action'. As our system passes from Solon to Pericles, to Alexander, to the Caesars, a 'recovery' vehicle emerges in halting steps from the Judaic branch as the rising oikoumene inherits of the benefits of parallelism.

It is significant, as a lost strain of this transition, that the tale of the Exodus myth expresses one of the first appearances in world history of the type of 'revolutionary ideology', however seminal in form. The Post-Exilic world was many things, and one aspect of it was a conservative continuation of the type of 'temple culture' already very ancient in the Middle East. The 'revolution' is still the 'revolution of the ages' with its transparent symbolism of 'new age' and 'Egyptian repression'.

2.7.3 Aryans, Hinduism, and a Buddhist Revolution

The history of classical antiquity in the occident is a braiding of Athens, Jerusalem…and Benares. Beside Israel stands the mysterious India, the great foundry of religious consciousness in the history of civilization. The source of this contribution, we suspect, is very ancient, already so by the time of the emergence of Buddhism, which is a kind of reform movement, and baton transfer from the Jain tradition.

The Primordial Tradition It is incorrect to see the source of Indian religion in the Axial Age. The primordial 'Shaivism', the source of yoga/tantra, probably appears in the Neolithic period. The question of Indo-European invasion/migration has muddled the whole history with a confusing 'something' called 'Hinduism' and its Vedic interpolation. Note the further comic irony that the (spurious, no doubt) periodization of the 'dread Kali Yuga' puts the classic era of the Axial period, Hinduism included, in the rubric of decline!

Shiva and Dionysus Is much of what we see in the classical era a set of remnants from an earlier period of Neolithic religion, spread across an entire oikoumene from India to Europe? The thesis is plausible in the abstract, while the details remain controversial.[16]

Fig. 2.39 Jain manuscript

Both Israel and India are considered 'spiritual cultures', but this prejudicial notion does not correspond to the real facts, and if we observe carefully, and then consider first China, and then Greece, we will see a spectrum, not a dualistic division. In fact the Axial period of India shows a remarkable resemblance to the Greek and Judaic cases combined, a system of city states suddenly crystallizing a tradition in a spectrum of philosophers and sages. The emergence of Hinduism is deceptive, for it is a hybrid created between the more ancient, probably Dravidian, tradition, and the peoples of the Aryan invasion.[17]

The history of India, and of its religions, can be very confusing in this regard, due in part to the cultural contradictions of its different traditions. The question of the Aryan invasion has produced a set of attempts to deny the reality of that process whereby an Indo-European migration resulted in a hybrid cultural formation of the Aryan and Dravidian elements. The grafting of the Aryan rule of caste on a religious tradition in which it was absent creates the distorted phenomenon of Brahmanism, and a subtle exploitative field of guruism.

---

16 Alain Danielou, *Gods Of Love And Ecstasy: The Traditions of Shiva and Dionysus* (Rochester, Vermont: Inner Traditions, 1984). The works of Danielou contain a clue in plain sight to the confusions of Indian religious history, but must be taken with caution.
17 Diana Eck, *Banaras: City of Light* (Princeton: Princeton University Press, 1982).

A Riddle Resolved 157

Stream and Sequence: Buddhism The case of Buddhism in India is spectacular, and a classic case of our stream and sequence effect. The streams of primordial Shaivism and Jainism are sifted and refined to produce a world religion ready to ship outwards in parallel to Occidental monotheism. The streamlined Buddhism carries none of the baggage that will chaotify so-called Hinduism.

Dates of Buddha There is a considerable effort to revise the dates of the Buddha. This is quite suspicious, although a later date would in some ways conform better to our thesis: the seminal era of Axial innovations is followed (as with Ezra and Nehemiah in Israel) by a codification of a world religion.

Post-Axial Shaivite Revival The stream and sequence argument can help to sort out post-Axial Indian history, for the resurfacing of the primordial Shaivism generates a series of indirect effects that can be confusing, for example, the sudden odd appearance of 'tantra' in a Buddhist context.[18]

Many commentators, and critics of the Aryan invasion hypothesis, have pointed to the great antiquity of Indian religion. But this is not an argument against the relatively late appearance of the Indo-Europeans, merely a suggestion that earlier, perhaps the Dravidian, cultures were the primordial vehicle of the ancient from which the core of Indian religion sprang. Once seen in this light, many of the problems that distract us from a correct picture of Indian history fall away. Beside this lies the tradition of Jainism, which seems to come to an end in the Axial Age, even as it spawns a successor tradition in the emergence of Buddhism. We must note the apt application of our 'stream and sequence' argument, and the way in which, through all the confusions, the Axial period seems to resolve the stream by creating an element of Indian religion for the sequence, by creating a global vehicle, Buddhism.[19]

Thus India, if we care to set aside our western viewpoint, shows us something preserved from great antiquity, and it would seem that we have glimpses of the birth of the great religions in the Neolithic. In any case, the primordial 'religion' of Shaivism, from which springs the lore of yoga and

---

18 Alain Danielou, *Shiva and the Primordial Tradition* (Rochester, Vermont: Inner Traditions, 2003), cf. Chapter 2, "The Shaivite Revival From the Third To the Tenth Centuries C.E.".

19 Alain Danielou, trans. Kenneth Hurry, *A Brief History Of India* (Rochester, Vermont: Inner Traditions, 2003).

tantra, lurks behind the later results that we see in Hindusim and Buddhism. *Before* the emergence of monotheism, the impulse of the sacred was preparing to leap beyond the notions of the transcendental or the conceptions of

Fig. 240 The painting was done in Han Dynasty. It depicts the story that Confucius talking with Lao Tzu who is the founder of philosophical Taoism in Spring and Autumn period.

divinity to base religion on inquiry into consciousness.

The tendency of Westerners to see a single linear track of civilization, the 'rise of the West', and forgets that the modern transition in its sudden unbalancing westward of the eonic sequence, is a very recent phenomenon in a once relatively backward zone of world civilization. It is almost impossible to sort out the emergence of, and relationships between, the forms of the classic yogas as they appear already before the Aryan entry into India, and reappear blended with Vedism and its issues of sacrifice, polytheism, and caste in the later Hinduism. The sudden eruption of Jainism and Buddhism, in period, is a clue to the later loss of the correct picture.

The earliest period of Indian history has already seen the civilization of the Indus come and go as the entry of the Vedic Aryans finds their religious culture to be typical of the proto-Iranian, and proto-Germanic spiritual cultures and the elements of the divisions into castes that are still visible in some aspects of Greek and Roman culture. The mystery is where the elements of the great yogas come from if not from the Vedic culture that shows a completely different character. Already these elements are visible in the famous cylinder seal of the meditating yogi found in the Indus

A Riddle Resolved

The pattern given by the Axial Age (we should anticipate the later results) only makes sense in a larger context which we can summarize briefly, and this case help to resolve the remarkable Chinese case…

Our system appears to be a minimax problem in the context of trying to set a mainline (using stepping stone frontier effects) and at the same time spreading outward in parallel as in the Axial period.

1. each transition generates a diffusion zone

2. the successions follow a stream and sequence effect

3. the mainline resumes in a frontier area in the next phase…

We see a clear mainline in the 'core Sumer' zone, suddenly beset with the parallel mini-Axial Age with Egypt in parallel with Sumer. Was there a set of effects here in China and India, as in the later Axial Age. That seems unlikely, but we don't know. Thus they don't enter the transformation interval around 3000 BCE. But China and India are complex sidewinders with their own complex stream histories. They are also in the diffusion zone of Sumer. In this situation we see 'mideonic starts' in the stream phase, in part due to diffusion: in China (another classic case if Minoan/Mycenaean Greece) the result is a set of formations like the Shang. Then, although it is hard to decipher, we see the sudden entry of China into the macro system with a transition in the period from 900 BCE to about 400 BCE…It is hard to locate this full transition, but the absolutely classic clue is seen in the on-time appearance of Confucius, Lao-tse around 600 BCE. Similar remarks are clear in the Indian case.

archaeological nexus. A considerable revisionist literature is now challenging the standard version of the Aryan invasion. But the picture is still unclear.

Upanishad It is almost impossible to grasp the complexity of Indian religious history without seeing the context of the eonic effect, or the Axial Age. The sudden appearance of the Upanishads in the exact time-frame of the transition, morphing out of quite different elements, is one of the most remarkable emergent processes of the transition. The transformation does its job, even if the result is misleading, i.e. it seems the outcome of some kind of Aryan Vedism. But in fact it is a primordial tradition picked up in the field of the eonic transition.

Fig. 2.41 Christian persecution Note that Christianity, Islam, and Mahayana arise outside the Axial period

Jainism It is Jainism that is carrying the great tradition of yoga from an earlier age, and these elements flow into the timely recreation of that tradition in Buddhism, and then in so-called Hinduism. The figure of Parshvadeva, a Jain teerthankar in the eighth century BCE suggests that a seminal transition now almost invisible to us was the decisive action in the gestation of the later Hindu and Buddhist outcomes.[20]

For our account, we can remain neutral, but the eonic context clarifies at once the way in which Buddhism suddenly appears in still another example of the 'relative transform' effect applied to an incoming stream, taking a bird's eye view over millennia. In essence, and in exactly the same time frame, we see localized cultural elements turn into a global religion rendered independent of cultural context. By the time of Ashoka we see the same passage to 'oikoumene integrator' in the early mixed forms that are characteristic of the Persian Empire. This eonic isomorphism with the Judaic case is entirely remarkable, and explains why Buddhism seems to stand out from its Hindu background. The great Hindu comeback against the Axial

---

20 Danielou, op. cit., pp. 32-35.

A Riddle Resolved

Fig. 2.42 New York Skyscrapers

Buddhist 'revolution' produces the world of the misleading *Bhagavad Gita*.

The emergence of Buddhism in the standard accounts is just after our divide, ca. -600. Some scholars now put this date forward, which would be appropriate also, since we can see that Buddhism is appearing about the time of the Ezra era in Israel. Our actual transitional era is almost lost to us, in detail, and produces the sources of the remarkable *Samkhya*, and a great deal more in a great flowering. All this is almost perfectly matched to our eonic model, which should allow us to stand back and put this era in perspective. Please note the appearance of another classic example of the relative transform (of a religion) that we have seen already in the steps of the eonic sequence. That is, the stream of Indian history already contains what the Axial Age will amplify and turn into the exteriorizing world religion of Buddhism. We should note, however, that 'Hinduism' in the post-Axial period is essentially still another relative transform of itself, and thus on its own terms an 'eonic emergent'.

The interruption of the rationalistic Buddhism between Vedism and the later Hinduism is the giveaway, however indirect, of the redirected stream so evident in the synchronous world of Israel and Greece.

As Prem Nath Bazaz notes in *The Role of the Bhagavad Gita in Indian History*:

> The seventh and sixth centuries B.C. witnessed in India, as in Greece, an intellectual ferment. Dissatisfaction with the Vedic natural religion

A Rebirth of Freedom…Cycle, System Return…

We are set to leapfrog into the future. We can note here the frontier effect about to occur as Europe is seeded and the Roman World expands to its limit in the European sector, the source of the next advance, almost precisely at the limits of expansion. It seems like there is a distinct 'Western Civilization' that is in some fashion doing one history but that is an illusion of perspective. At this period Europe is a backward fringe area in the sequential zone of the later Roman system. As such it begins to receive, finally, the rich influences of the eonic sequence indirectly. It rises from its slumber slowly but surely. Europe will be the last frontier diffusion zone left in the Eurasian field, Japan being another such. But Europe is fortunate in so far as its medley of tradition will inherit the output of two transit areas, the Judaic, and the Greek, and its languages are a closer match to those traditions, facilitating the spread of the Axial novelties.

The suggestion of the eonic sequence is return on the far future, and we are already in the modern period, as we find its seeds as much in the dilemma of the Hellenistic, as in the economic derivations of capitalism from Medieval Christendom. We have come to another 'what next?' point. And we already know the answer, and, further, see why students of the early modern are condemned to equivocate the causality of the European resurgence. The modern period is gestating just here, for system return after 2400 years in a jump diffusion zone, i.e. at the fringes of the tide of expansion. There will be few candidates. The Hellenistic passes into the Roman Empire, thence at the boundary in Northern Europe we find a zone both fed the great advance, and yet still virtually untouched. Granting the dangers of 'discrete oversimplification' as against the sterility of 'continuity models', we are nonetheless drawn to the strange conclusion that the rise of the modern shows 'system return' in frequency, in a jump diffusion zone, as the 'emergence zone', this time unique, for the great roll of eonic sequence rolling out of the Neolithic.

gave rise to speculations about the origin of the universe and things contained in it...There arose early in the sixth century B.C. an order of *paribrajakas* (literally 'wanderers') who were intellectuals devoted to search after truth...The movement of *paribrajakas* spread far and wide in Northern India; they were accepted as harbingers of a new age...[21]

The views expressed in this flawed and highly charged but useful book suggest the fact that Buddha was not only a religious founder, but a social revolutionary, a view with a bit of its own myth perhaps, but the account gives an apt descant on the Axial period compared with the later destruction of Buddhist India. It is time for some fact checks on all accounts until the record is straight. The stage of the *Bhagavad Gita* represents the reactionary phase of Neo-Brahmanism that came later. This history deserves an account by a modern leftist, and may cure our contemporary New Agers of sentimental views of the history of guruism.

East and West? There is no 'philosophic' East and West, although over time a kind of misleading differentiation arises. Those who find a something called 'Western civilization' are really speaking about an artificial construct built around two transitions, whose final effect is a transmission of this mainline out of Sumer back onto the full Eurasian field. The mutual influence of East and West is continual throughout the classical era. Thus, many are the speculations about the interactive influences, viz. the influence of Buddhism on Jesus. We can hardly spot the exact blends, yet we can easily discover the overlap in the Indian, Judaic-Persian, and Greek-Roman cones of diffusion.

Lokayata The Upanishadic age was a close cousin, that is, temporal parallel, of the world of the Pre-Socratics and Sophists, and its spirit was extraordinarily broad, and in many ways deeper. Jawaharlal Nehru's *The Discovery of India* describes the contemporary rescue of over fifty thousand Sanskrit manuscripts on what, given the extensive destruction, must have been the great quantity of ancient literature. "Among the books that have been lost is the entire literature on materialism which followed the period of the early Upanishads." This is the lost world of the 'lokayata', reflected in the *Samkhya*. We have become so conditioned to the 'material'/ 'spiritual' distinction that we can barely appreciate the way the realm of religion was once cast (among a spectrum of such) as a naturalistic philosophy.

Quest for the Historical Gita The history of Indian religion is a highly difficult swamp laced with the propaganda of the Hindu

---

21 Prem Nath Bazaz, *The Role of the Bhagavad Gita in Indian History* (New Delhi: Sterling, 1975), p. 82.

reaction to Buddhism. *The Gita As It Was, Rediscovering the Original Bhagavadgita*, by Phulgenda Sinha, attempts to uncover the text of the original non-theistic *Gita* from the layers of distorted interpolation that brought it to its present state. The idea of a Buddhist revolution is partly an anachronism, but we do see in the contrast of Buddhism and Hinduism another smoking gun example of an 'eonic effect'.

Fig. 2.43 Sumer

An Evolutionary Psychology: Classical *Samkhya* The legacy of ancient *Samkhya* with its universal naturalism might prove of help in a period of extreme reductionist materialism. Charged with materialism *Samkhya* is then again charged with idealism, and dualism, and shows a remarkable collation of opposites, and a distant resemblance to Kantian thinking. One problem is that this discourse has already been appropriated for any number of metaphysical speculations about cosmic involution, which don't do justice to the original. At the point where it appears in the *Bhagavad Gita* it has already lost its original significance. The world of *Samkhya* points in principle to everything known in the ancient sutras, and this material is late in terms of our eonic Axial period, but still close to its source.

The history of Indian philosophy seems determined to place a Kapila right on schedule as an eonic sage, as the creator of *Samkhya* in the time-period 600 BCE, as though to assist our delineation of eonic architecture. The evidence suggests that it was emerging from an Upanishadic phase that is registered even in the *Mahabharata*. The exact form that it took in the age of Gautama is not clear, but the influence on Buddhism is so obvious that we can feel confident that the main features of the system were more or less in place in the time of Buddha. This is slightly out of character in the Upanishadic context, as these progress into the consolidation of Hinduism, but we should note that the whole tradition here has never truly been shown to have anything to do with Indo-European, or Vedic, religious traditions.

The fate of this system was denunciation by the later Shankarans who had quietly expropriated its terminology and concepts, witness the references in the misleading *Bhagavad Gita*. And they were not the last. Great later embarrassment rings through the history of mysticism and religion in the fact that the great breakthrough of the classical Indian transition produced

A Riddle Resolved

> **Revolutions Per Second: The Rebirth of Democracy**
>
> The onset of the French Revolution deserves as much as any date in history, beside the more glorious flagship American onset from 1776, the importance that has risen around it, as the period that initiated a shockwave of modernizing change that was national, then continental, and then global in nature, and whose cornucopia of diffusing consequences is still with us. That it was directly influenced from the fringe by the American revolution in its virgin open spaces is itself significant, and it was therefore a subtle recursion, in the broadest sense, of the experience of the English Civil War, and its aftermath, the Glorious Revolution of 1688, against a backdrop of the rising liberalism and deeper underground radicalism generated from the philosophic, scientific, and revolutionary experiences of the English... WHEE

Despite its ragged character our model accurately reflects some stunning properties: the final phase of revolution climaxes near the divide (ca. 1880 or 1750-1850) and this reflects the interaction of 'system action' and 'free action'. The macro can't fully create freedom through system action, and free action is often powerless. Thus, system action creates a potential, then at the divide the transition to free action puts freedom realization in the context of autonomous self-creation.

We see exactly the same effect in Greek antiquity near the suspected divide period ca. 600 BCE and the remarkable appearance of Solon initiating the somewhat disorderly creation of Athenian democracy...

a 'materialist' mysticism. But such a thing was quite natural in the age of Buddha and Mahavir, although we cannot say what the true original form of all this was, for the Shiva cult and its yogi far predate Buddhism. All we see now are the later redactions of the Hindu medieval period, so concerned

Fig. 2.44 The Slave Market

under the influence of Islam to conceal the whole subject in a monotheistic wrapper.

The sutra posits a dualistic distinction of *prakriti* and *purusha*. This double aspect model is the key. The 'spiritual' principle is strictly segregated from the sources of natural manifestation, and these include mind and soul. The 'spirit' of man is higher 'material', and not the same as *purusha*, which is uncreated, and uncreating. *Prakriti* comes in two aspects, uncreated, created. It is this unmanifest prakriti that is the obstacle to easy self-realization. The value of the *Samkhya* approach is to see that one mistakes one's spirituality for what is in reality a material manifestation in subtle form. The beauty of the system of *Samkhya*, the codified echo of some unknown Buddha, as ancient as the speculations of Thales and as deserving of a place in the Smithsonian of proto-science, is its consistency and simplicity: everything is 'material' in an all-encompassing naturalism, that is, all is of a piece, matter, energy, mind, purpose, god, and yet beside this is a witness, perhaps misunderstood as 'consciousness', a term they did not use, and which

A Riddle Resolved

mis-portrays the element '*purusha*'. It is misleading indeed to translate the term '*purusha*' as *consciousness*. This 'dualism' then receives a sort of myth of the relation of the two in a striking image of a kind of evolution as punctuated equilibrium. This witness does nothing, and is neither god nor creator. Everything comes into existence from primordial matter as a cascade of evolutionary triads or gunas, doubling in number in some later formulations: 3, 6, 12, 24, 48,... This aspect is speculative and has degenerated into its own form of bogus cosmic mechanics that found its final burial grounds in the pastiche of such as Ouspensky.

Fig. 2.45 "North Against South"

The dualism of 'spirit' and 'matter' disappears and becomes a 'dialectic' or triad, in a tetrad including *purusha*. It is not a dualism of matter and spirit, but a dualism between the 'unnamable, but named, *purusha* and a natural triad, of three 'matters'. Some of these 'matters' are unmanifest, and that's what causes the confusion of spiritual *samsara*. The point is that the higher range of this triad, the 'sattwic' is confused with the spiritual. Perhaps it is the spiritual, but there is something beyond that. This dialectic is biophysical, the fact of the body, the mind, and the triadic 'connector', 'e-motion', desire, etc,... Science might have grown better in this acidic soil, as it thrashes about in Cartesian schizophrenia (although Descartes is attempting a similar gesture), sinking deeper even as Descartes is denounced, unable to get its 'materialism' in order. *Samkhya* is one great key to the labyrinth of Indian spirituality, tracing its origins to the era of Buddha.

> *Samkhya* can be useful as a reminder that religions are not spiritual but upsurges in *prakriti*. Yogis hitchhike on the form and one day are found to have slipped away as the *purusha* element, allergic to religion, subtracts their name from the religious roll call. We see the point looking at the eonic effect with its ambiguous, now material, now spiritual, eonic emergents. The distinction of matter and spirit in

Western language tends to divide the 'sattwic' from the whole man to call that the spiritual.[22]

2.7.4 Axial China: Continuity and Discontinuity

As we see from the parallel echoes in this synchronous phase, there is no inherent difference between the East and the West. The Chinese Axial intersection is beguiling because its isolation shows the eonic effect in a displaced and attenuated form, and the effect of a creative period one third of the way through an otherwise relatively continuous stream.

> **The Chinese Axial Interval** The strange thing about the Chinese instance is that it is almost invisible, on the surface. But the clues are there to an exact match if we can understand them. The change of character in the eighth century Chou era, the appearance of classic tradition ca. -600, and the resolution to empire in exact concert with the Hellenistic, tell us that we are seeing something in disguise, or else a politico-democratic trend toward equalization ideology that never fully realized itself.

The Chinese case proceeds rapidly toward integration as empire, as a political construct, after the Warring states period, in the same timeframe as the Hellenistic. This continuity is remarkable and we find the later Sung period, and the near take-off of a great economy where the West is in a medieval period. Part of the difference lies in the relative isolation of Chinese civilization from the Western transitions (although not from external invaders). However, the diffusing sources from the first transitions in the Sumerian field are what trigger (as far as we can tell) the rise of the mideonic Shang era, and before. Note by comparison the immense number of collisions in the Mesopotamian downfield, resulting in the emergence of the integrator religions. Taoism and Confucianism are the parallel equivalents, a unique blend of the political, philosophical, and mystical. There is an irony in the later diffusion of Buddhism to China, for in Taoism we see another variant of the same.

What evolutionary theory will then accept a transition one third of the way through its history? Thus, as we ponder the relevant era in light of this continuity, our consideration of the fundamental unit of historical analysis will force us to consider something operating independently of the actual

---

22 *Classical Samkhya, An Interpretation of its History and Meaning* (1979), Gerald Larson.

stream combinations of culture. Is there any support for such a strange idea in the literature? Kwang-Chih Kwang, in *The Archaeology of Ancient China* notes the turning point in the Chou era (eighth century), and observes, "A new era in the history of North China began in the Eastern Chou. In political history, ancient China consisted of the Shang and Chou dynasties, but in cultural history, the subdivision may be placed at the Middle of the Chou dynasty, dividing the Shang-Chou periods into two stages." [23]

Fig. 2.46 Gothic Cathedral

Far too much analysis has been given to the question of why science in the modern sense didn't emerge in China. Despite being a very advanced culture able to develop in isolation (though, please note, with nothing like the emergentist democracy phenomenon), the emergence of modern science appeared in a less developed region. But as we look at the eonic sequence, the reason is clear. The mainline eonic sequence tends to hug its basic center of gravity, and diffusion rich fields near that.

Comparing the Chinese and Greek transitions is interesting because of the clear, but intangible, common denominator behind the clear difference in historical generation, and the ringing chord of philosophic 'enlightenment'

---

23 Kwang-Chih Kwang, *The Archaeology of Ancient China* (New Haven: Yale University Press, 1977), p. 386. A developmental history of the *Analects*, Bruce Brooks & A. Taeko Brooks, *The Original Analects* (New York: Columbia University Press, 1988).

that comes ashore in spite of causal diversity. The history of its transition is the history of its philosophic generation, and the transposition of 'science, mysticism, monotheism, philosophy, and political ideology' in recombination that shows a glimpse of the 'eonic abstraction' at work. In the strange dynamism of the Taoism and Confucianism we find the synchronous 'eonic

Fig. 2.47 An Indian Mosque

equivalent' of the occidental monotheisms, an extraordinary alternate universe that bypasses so many of the confusions that arise in the west, and a clear indication that the forms of 'revelation' are in fact 'free action'. But the western religious forms will end better adapted to cultural integration, at least in principle. In practice, the entry of the Chinese philosophies into the West almost from the beginning of the modern era and their popularity and influence on the *philosophes* shows the real case of greater universality.

> **Science and Civilization in China** The example of China is instructive, since it is so lateral to the center of gravity of eonic sequence, yet shows uncommon continuity, along with technical expertise that never, however, gets the full 'eonic amplification' of the emergent science all too obviously hugging the 'central track' out of Sumer. The recurrent birth of science is a function of the triple phase track out of Sumer, with the mideonic efforts to keep it afloat the gestating result by the Islamic world during the medieval slump. Even so we find the invention of

printing, gunpowder, and the compass as mideonic Chinese inventions that dawdle in isolation to first cross a transition after diffusion to the stepping stone region in the West. The attempts of Joseph Needham to study emergent science in China are perhaps excessively focused on the wrong factors. The main issue, given the 'case of the missing centuries', is the center of gravity of the eonic sequence, not the claims of Western technical superiority. China never even received the main early scientific texts, or had the direct influence of the Ionian or other intimations much more available to the 'near-far' Milesians. We see the clear difference of technostream and the intangible eonic determination.

2.8 On the Threshold of World Civilization

The great era of world transformation passes, and by -400 we can see the waning of the effect. The outside date, -200, for Jaspers' Axial Age is far too late. By then the Athenian world is gone, the Roman Republic is beginning to suffer strains, and era of Empire is soon to come. The great religions are coming into being. We can see the difference in the post-transitional period at once in the passage of the Greek world to the Hellenistic Age. In Greece, the difference is dramatic, visible by the fourth century. *Polis* is turning into *cosmopolis*. Indeed it was in this period, as the classicist H. Kitto notes in an essay on the decline of the Greek *polis*, that the word itself, '*cosmopolis*', was coined to serve the passage to an allegiance to the greater community of man. A great expenditure of history grew from this point to prepare a first universal cosmopolitanism.

In *The Harvest Of Hellenism*, F. E. Peters opens his depiction of the great oikoumene that is unfolding by noting, "This is a book about a second generation', the first generation being that of the Hellenes from Homer to Aristotle, the second one 'without a name', Greeks, Macedonians, Romans, Syrians, Jews, Egyptians. They came "under the spell of the Hellenes… condemned or blessed to reap where their spiritual fathers had sown."

In fact, Plato and Aristotle are a bit late, but show the last consolidation of our transition, before the rapid waning of the eonic dynamic. The period of the transition from the classical flowering to the Hellenistic world is the most solid, and the most confusing, period where the evidence of historical directionality, and a mysterious misdirection, come together. One aspect of the change is evidenced in the neo-authoritarianism of Plato denounced by Popper and can be found in the minor classic, The Liberal Temper in Greek Politics, by Eric Havelock. The use of the term 'liberal' for the Classical Greeks will not work. However, the basic point that Havelock is making is valid, by any terminology, in showing the change of character that came over the Greek world in the generation of Plato. The Sophists are maligned, but they are exemplars of the inchoate transition figures.

Our eonic model shows us at a glance the psychology of religion that arises in the Christian world, and the compulsion men had to think there were spiritual forces operating on their future, generated from the transition. They were correct, and correctly produced a myth of the eonic effect! But it is not the action of divinity. Only secular thought can summon the brusqueness to remind his religious brethren that a divinity would never act according to the hopelessly confused outcomes of monotheism, as the mideonic stream jackknifes and produces antisemitism, and the rival emergent teleological vehicles struggling with medieval inertia.

Fig. 48 Wilberforce

The world into which the transition passes is one aspect of the perception of cycles that can do harm to progressive advance. As the sociologist Krishan Kumar notes in *Prophecy and Progress*,

the backward-looking spell of the memory of the world of classical antiquity remained, to bewitch thinkers into a sense that the great, golden age of man was really in the past, by comparison with which present times were mean and secondhand. This spell was decisively broken only towards the end of the seventeenth century.

Our framework now highlights the great historical drama of 'decline and fall', the progression toward religion and empire as oikoumene generators that will characterize the immense interval, the mideonic period, from the end of the Axial Age to the rise of modernity.

> **Decline and Fall** The succession to the Axial Age provides us with an awesome display, and partial explanation, of the mechanics of 'decline and fall', and in the Occident the final collapse of the Roman Empire about a millennium after the onset of the 'new age' is the demarcation point for the tellingly named 'middle ages'. We should be careful to distinguish the mechanics of our eonic effect, as self-organization, from declines of civilizations, which are due to other processes. This pattern is the mirror image to the eonic sequence, and is often the source of comparisons for critics of modernity. But the two situations

are quite different. Please note that there is no inherent inevitability for this mideonic decline. It is possible for the system to advance from its transitional periods, and do that consistently. But we can see how the logic of disorganization slowly overtakes the larger system created by our eonic sequence, and this requires 'restarting' at the point of the next cycle. A frequent comparison of modernity, or else the 'American Empire', to the decline of Rome enters into an ideological sermonizing against the imperialistic capitalisms of modern nation-states. But these comparisons are misleading. Even if we accept the possibility of comparison of such different eras and cultures, the modern system would still be at about the point corresponding to -400, with almost a millennium to go! The decline of the Roman Republic into Empire, and of the Empire into medievalism are two separate things.

2.49 Roman ruins

2.8.1 Slavery, Abolition, and Eonic Sequence

Classical civilization is reaching a crisis point here in the Roman world, beyond which no progress is possible short of abolition, which, please note, ignites explosively just at our next divide.

Consider antiquity, then, in the wake of the Axial period, then the beginning of civilization. A system set to advance, with new elements of economy, simply nosedives, the factor of slave society growing progressively worse—until the medieval period, in the West at least. Christianity and Islam get honorable mention here, but they simply were unable to solve the problem, however much they laid the foundations for a 'New Man' able to handle the elements of modern civilization. We cannot neglect their crucial seminal contribution, nor blind our eyes to their inability to resolve the problem in full. This factor of slavery exists from the beginning, but never as a true functionality of real civilization, which cannot come into existence in such form, we should think.

In the worlds of Sumer and Egypt, the issue was ambiguous, but slowly deteriorating. But Marx is right in one way, the factor of 'implicit class struggle' attends the birth of the state. Critics of Marx correctly point out that 'class struggle' never appears until modern times. But that misses the point. The dilemma arises from the nature of the state itself, implicitly. One should wager a sum we would see, with close evidence, no intrinsic slavery at those points where state-emergence shows eonic determination.

It should be, we suspect, like the Greek case where the new and future mode is stillborn in the midst of the old. We can't be sure without facts. After all, the myth of Exodus clearly records a great drama of 'class struggle' and incipient revolution. But we need better historical evidence. Slavery has perhaps existed since the Paleolithic in some form. And it seems as if 'history' is compromising here, 'to get things done', until the rise of industrial civilization and abolition. We simply can't make that assumption so easily. A discrete-continuous system simply resets itself in a new future, and the past is truncated.

The subject, peasant, Neolithic farmer, or embryonic citizen, as an entity of socialization at the beginning of civilization, might be exploited, but he is an embryonic 'citizen', even before the grandeur of the Pharaohs. Class struggle is thus implicit in the birth of the state. But as to slavery, we might speculate that the system is inchoate and can go either way. Freedom is born in parallel with amplifying slavery. Thus we have no real evidence that slavery shows direct eonic determination. The point is that we cannot assume that 'Big History' is exploiting slavery on its way to a better future. Our transitions simply happen, the idea of freedom emerges, doesn't take the first time, and the result is history getting worse, not better. But there will come an end state to the tragic era of slavery, but it will come in eonic time, not by slow evolution of liberty!

We could just as well say that men in the direct line of the eonic sequence prove unable to realize its real direction, or mix elements outstanding to the mideonic realization. Cynical Machiavellians might take note of just how much of humanity's time they have frittered away. The Roman world can go no further, so to speak, until the issue of slavery is resolved.

All this may seem to be naïve idealism, but it is a reminder that we can specify no active agent behind our eonic sequence, which becomes 'active' (?) briefly, shuts down, and waits, apparently. But we do see something more like Santa Claus dropping gifts at regular intervals than some bloodthirsty spirit moving toward the 'end of history'. It is savage man, projecting his carnivorous instincts against the universe that seems to be the problem. In general, while a realist attitude toward slavery

A Riddle Resolved

might seem the normal view, world history appears to mostly a legacy of abnormalities, so far. The point of our argument is to summon up a dialectical antithesis, and then demand hard proof in a deductive model of any proposition asserting the 'stage of history inevitability' of slavery.

> **Market Evolution** It is here in this period that the idea of the evolution of the 'market order' as the basis of historical sociology will fail: it does not evolve spontaneously against slavery (although the Roman Empire, it could be argued, has a considerable market evolution based on slavery). Instead the whole western system peters out and ends up in a Christian/Islamic medievalism. The picture of civilization at this point was not pretty, precisely because the market order was too immature to pass beyond slavery. The great irony, for those who think 'self-interest' as secular religion can explain history is the long delay in the birth of (modern-style) capitalism and it almost seems like there was a need for a long religious preparation. The market order requires sophisticated help like everything else. We still see the last phase of the confusion in the modern transition where freedom grows in relation to the core, while slavery is exploited at the fringe, resulting in the historical confusion of the American paradox, a slave state grafted onto democratic generation. The ancient system never achieved the market order as it amplified the slave system into such institutions as the Roman *latifundia*. Such statements require the obvious qualification and challenge of noting that capitalism was essentially already born in one sense, in the snafu over 'relative transformations' our model handles properly.

2.8.2 Religion and Empire

In context of the eonic effect the generation of Christianity and later Islam (and Judaism as we know it now) from the Israelite core phase suddenly falls into place in our explanation. The mechanics of these religions is impossible to understand without an eonic model, that is the distinction of System Action and Free Action. The action of the large-scale historical component (which call evolutionary) is one thing, its realization by men, Free Action, is quite another. Many of the endless confusions over religion will be clearer if we understand this difference. And one consequence is that, according to our rules at least, we cannot explain the mideonic religions to come, i.e. our system does not control the coming mideonic futures, although these are sequentially related to some core potential in the transition, and the Old and New Testaments of the Christians virtually say just that as they create an eonic myth of the mysterious system they find themselves in. It is easy to fall into a ditch here, and it is good to be wary. It is helpful also to look at the Buddhist example for comparison to see the strange core process at work. But we can see how the general pattern is in some fashion latent in the transitional period.

> **Christianity:** A mideonic phenomenon It is important to remember we are dealing with eonic history, and this does not produce an all-inclusive account of its mideonic periods. It is not our job to fill the blanks with some simplistic account of, for example, the emergence of Christianity which is not a part of our eonic sequence. But our schema produces an exact, but abstract, rendition of the emergence of Christianity (or Mahayana, or Islam), and then comes to a stop, our job done, as it were: the transitions produce a seed material as macro-action and these proceed toward the diffusion field there to generate materials for the generation of an oikoumene in the field of micro-action. It is hard to think of a better (eonic) portrait of the emergence of Christianity. But even as it explains, it explains nothing, which is as it should be. These mideonic religions are creations of men, not the eonic sequence, expressions of their freedom under eonic determination, or macro-action.

Thus, it is very easy to produce a plausible scenario of the way our model 'generates' the seeds for what comes, as long as we are wary of thinking we can grind out the particulars with eonic analysis, we can't. The case of Christianity, for example, is both exceedingly obscure and completely transparent, at least with respect to our model. The Judaic stream brims and overflows, as we see a spiritual movement suffer the strains of transcultural integration and break away into a new religion.

These religions are now challenged in the next phase of our system, and the New Age effect is starting over. Nothing in our account requires any future for religion, since this category tends to the ad hoc of its age period. But modern secular thought can barely do justice to the immense task performed by the era of these mighty oikoumene integrators whose impulse moved toward the protection of disparate peoples and diverse evolutionary groups. Secular would replace this integrator theme with Darwinian thinking, then wonders fundamentalism is resurgent. We are so distracted by the metaphysical issues of theology that we fail to see the gestation of a new man from the action of these mysteriously emerging formations rising to challenge, then defeated by, the world of empire.

The critique of someone like Nietzsche of the onset of these champions of spiritual equality is unfair, and historically blind, and we must dread a future constructed of scientism, Darwinist reductionism, and neo-barbarism if an improper or ill-considered exit into secularism entirely displaces the impulse toward the community of man these vehicles created. Modern man must surpass these religions without regression. Our modern transition has already laid the foundation for a resolution of these questions. But we must note the way that these mideonic periods tend to fall into chaotification. Darwinism will almost certainly reignite an 'Athens-Jerusalem' style collision if it grows to overtake the global consciousness. This won't have anything to do with a renewal of the ancient religions. A similar effect is very clear in the far left of the nineteenth century, a materialist movement.

We will remain within the deliberate restriction of our model and issue its stern reminder, that these religions are mideonic constructs. That means that men created them, and how they did that is simply not clear from the evidence, and requires some grounding in the more adept spiritualities of India. Especially with the birth of Christianity is that the case. It is a puzzle with too many missing pieces, one of them the charming tidbit of the 'three Magi'. The triple action of John the Baptist, Jesus and Paul is hard to reconstruct, and too coordinated to be chance, but too ad hoc to be divine action. We can easily suspect, but not prove, something missing is crucial. The story of Paul's conversion is a giveaway, but a giveaway to what? A true tour de force of concerted action whose choreographers we do not see, and whose tactics we may never know.

Thus, we can now see the era of phase pass into a distinctively different period of ecumenization, one that we can call 'mideonic', not really 'medieval' in the normal sense, or even in decline, but distinctly 'inside' the new boundary created by the era of phase. Comparison with the previous cycle tells us immediately, as one clue, what is afoot. In the Mesopotamian sphere, small starts rapidly degrade into Universal Empires as the false integrations of the ecumenizers, Sumer to Akkad. A new answer is needed, and the beautiful Greek world, passing to the Hellenistic, the Roman to follow, will prove unable to provide it. The world religions appear in the passage from phase, and the occidental monotheism will speak from Sinai in the myth of Moses, from a people, the effect is beautiful, whose Incredible Shrinking Kingdom actually disappears at the climax of transition! Nowhere at all touches the grand Void and spawns the Islamic chase toward the far-flung Everywhere as one in the Kingdom of...

The most obvious indication is the truly ominous decline of the entire system, in the West. The fall of the Roman Empire is the token piece here, yet we might assume that our system predicts this, or the argument requires it. Not true. This massive decline is not visible to anything like this degree in the world of China. Our subject is eonic rise, plateau, rise, not necessarily decline and fall. The problem is that system runs out of octane, and becomes humdrum, sluggish, then starts downhill slowly, unable to advance, among other reasons because of the factor of slavery. Our system might have had two

Fig. 50 Storming the Tuileries

millennia of democratic experiments. Instead modern man ends up doing tenth grade work in the eleventh grade. We see the drastic cutoff point, as the transition coughs, sputters, and dies across the board.

The eonic falloff phenomenon in the Hellenistic Age is the answer to Dodds' 'failure of nerve'. The difference is unmistakable very quickly, and proceeds from the era that started with the Iliad and passed toward that of Virgil's Aeneid. The contrast of Athens and Rome shows the clear difference of 'phase and sequential dependency indicated at the root of this analysis. We see the one blend into the other as the new era proceeds, and proceeds from the sturdy Roman Republic to the time of the Empire.

Mideonic Forces? The nature of our tale changes as we pass from eras of transition to the related sequential dependency of the mideonic world that arises from eonic generation. Our model has a problem, we can't explain the 'middle periods'. We designed it that way, on the basis of the evidence, the plus, beside the minus, of a discrete-continuous model. Everything in the mideonic interval defaults to 'free action', an apt and illuminating, though limited, approach, justified, however, by the facts. That is its value, and limit. By definition of our terms there are no such 'mideonic' forces, and the system proceeds on its own. And yet there seem

A Riddle Resolved

to be such. Something in the transitions generates the potential to create the mideonic realization. In one way the answer is right in front of us.

Our form of analysis creates a seeming paradox, the reverse of that of the transition. If we attribute 'driving force' to phase, and yet associate this with emergent freedom, we are confronted with the chance that after the phase we will wish to see 'unforced freedom' and yet more probably will find a loss of freedom. Such paradoxes are really a sign that we cannot apply conventional dynamical statements to the system we find. But the data reflects this feature of the model most definitely and we know we are on the right track to something because of such accurate reflections of the model. The terms of explanation are 'eonic determination' and 'free action'.

Sequential dependency is not determinism. Instead, information flows outward and there's a good chance the local future may conform to that information. The new influences of the transition diffuse outward, sort of hoping to influence the future, but more or less just keeping its fingers crossed. So what's to stop someone in the mideonic times and places to simply ignore the general 'evolutionary' direction. We see the elemental significance of religion as a core area generates a 'script of action', in the form of a corpus of materials, which, remarkably, even include claims on divinity saying, 'Do this', 'This is how free action' should behave. The tablets of the law, crystallizing as myth just at our divide, in the expanded abstraction of the state called a 'religion', flow outward into the field of free action.

> **Antisemitism, Mideonic Jackknife,** Teleological Tragedy One of the clearest indications, and liabilities of eonic evolution in our sense is the danger of jackknifing realizations in the mideonic period as the system action wanes just at the point where its productions meet a greater totality. The nature of our model allows us no use of the mechanics of transitions to explain the mideonic outcomes. And history reflects this, keeping in mind that our account of the Israelite transition is not theistic. We can see the difficulties and dangers of making teleological statements about the eonic mainline, and yet we tend to see the projection of the core transitions onto the greater field of culture as somehow the intended outcome of the whole process. The problem with this, and there are others, is that the middle period and the long term are different, and the result turns into a teleological ideology on the part of those realizing its action. The Jews and the Christians quite obviously diverged in their interpretations. This example should cure anyone of teleological thinking. We can see the quiet desperation of someone like Mohammed, 'start over from scratch'. The entire egregious and wrong result of Christianity with its Anti-Semitic strain is one of horrors of world history. We

The Enlightenment

With uncanny timing, the period of the Enlightenment climaxes at the end of the modern transition just before the point of the divide. This period is especially significant in our account since it is the last manifestation of the enigmatic macroevolution we have discovered, followed by the rapid shutdown of the eonic sequence in the next generation.

Although we associate the Enlightenment with the eighteenth century, its roots are really in the seventeenth century, and its true parentage still earlier in the era of the Reformation, as it rises to the Thirty Years War. There is a unity to the steps, from the breakdown of the Catholic world of theocracy, the partition of Protestantism, the ambiguity of authority followed by the disposition to reinvent the state or secure the elements of new sovereignties, Hobbes and the English War, in the 'bourgeois' economic and liberal mode of civil society, followed by the focus on the place of the individual discovered in freedom, to search for a new ethical self, and encountering the physics of the new materialism found from the rebirth of science as a system of the world. An almost timeless age in itself, and yet a moment in a larger sequence, the Enlightenment is seen best in its own context, which is its challenge to the past, more even than the future, as the birth of the idea of Progress bears witness to the rising breeze against doldrums of slow centuries. WHEE

should note that we see similar effects in India in the divergence of Buddhism and Hinduism and the long conflicts between the two.

Fig. 51 Tyndale

In any case, the confusion of Christians and Jews is especially tragic. It is logical, in retrospect, to see the transformation of the Judaic emergentism into a world religion as part of distributed evolution, but the actual details shows an arbitrary character, and a very dubious series of attempts to justify the result in theological terms. The modern period shows the whole danger all over again in the rise of the far left in the throes of globalization, and we need to try and find some resolution of the inexorable deviations of teleological claims on the future, owned by no one.

Christianity/Judaism, Islam, System Action, Free Action Our model produces a beautiful insight into the emergence of the great religions, so-called, but at the same time we must be clear that it takes a 'hands off' approach to their appearance since by the very nature of a discrete-continuous model they are beyond the range of our dynamical explanations, or explorations. They default to mideonic 'free action'. The most we can claim is that something in our eonic sequence, here the Axial interval, produced seeds that flowed into a diffusion zone thence to be raw materials for mideonic constructs, and the mediation of new oikoumenes. Full stop. And that much the evidence shows, most powerfully. And yet this 'explanation', even as it explains everything, explains nothing, and we must respect the historical integrity of these outcomes by opening a new file for their study. We must trace their historical chronicles without invoking the dynamics of the eonic sequence. Because of their occulted origins, that is extremely difficult to do. We have abstracted the question beyond the design argument visible in the Old Testament, and shown its eonic character, one the first Christians struggled with most directly. No designer would use a discrete-continuous action, it is clearly evolutionary, and makes sense in those terms.

The Axial interval of the Old Testament appears on schedule, while, for example, the initialization point of Christianity is given no explanation in our model. And that is right and proper. It defaults to mideonic microaction. All we can conclude is that later men, in the realization of the powerful corpus of eonic emergents appearing in the eonic interval, saw fit to do certain things that later became major religions. And they struggled even more specifically with their inchoate perceptions of an 'eonic effect' in action by noting the special character of their source point, calling that, misleadingly, an age of revelation, thinking further that certain prophets predicted what they were doing. This issue of prophecy confused them since we must doubt that interpretation, as we see that what occurred was at most a selective realization among a host of potential outcomes, the contrast of Christianity and Islam giving a powerful indication of this different potential realized.

Fig. 2.52 King James Bible, 1611

It might be that our eonic model is too basic, that a deeper dynamic is missing in our attempt to express the character of the eonic sequence. But we are bereft of the means to carry this further, although hints and intimations of such lurk in the data. For example the sudden appearance in concert of Mahayana and Christianity six centuries after the divide, both as schemata of redemption, must leave us wondering what we have missed. And the curious Zoroastrian character of Islam near the source points of that other tradition hint at a more complex picture than we have drawn. And the appearance of Sufistic traditions embedded in Islam shows us an experiment in 'religion-formation' taken to a very high level indeed, a phenomenon well beyond the capacity of our model to explain.

Let us note what later secularists tend to (wish to) forget, the theocratic ambitions of the great religions of the Axial Age, visible powerfully in the transmutation of the Israelite theocratic state religion into an oikoumene action script pool, leading to the projects of 'spiritualization of empire', however confused or unsuccessful the outcome, leading to the powerful dialectical reversal in the modern transition. This was a response to the degenerations of empire so obvious in the encounter of Israel and the Assyrians, for example. We need to take everything in its time, 'root for the team' in its time, and then do backflips as we pass to the successive stages of the eonic sequence.

Unfortunately great confusion has arisen in the emergence of secondary, often 'occult' or 'esoteric' spiritual traditions. We cannot rule out the possibility that emergent Christianity or Islam were the creations of historically undocumented agents of 'will' operating via proxies. The suspicious appearance of Sufistic agents in the background of Islam is one question mark. The previous appearance of such characters and their occulted feats has to be considered in the puzzling veil drawn in the New Testament around the basic chronicle, consider the beguiling appearance of the 'Three Magi', a sort of smoking gun of some kind. An ironic historical version of a design, human, all too human, argument lurks therefore in the attempt to decipher the undecipherable beginnings of Christianity. Whatever the case, what they did exploited the rich material appearing in the wake of the Israelitic transition. This well-tilled soil was a spectacular opportunity. They saw their opportunity, saw it as predicted by the Prophetic tradition and wove a new tapestry around that eonic saga, of which they were only partially aware. We can be almost certain these curiously veiled 'complots' lurk in the Buddhist sequence with their known ability to act beyond space via proxies. So everything about the onset of Christianity has to put into the category of 'unanswered questions'. The 'designers', whoever they were, leave only a cold trail.

It is significant that the eonic sequence operates at a deeper level than that achieved by Buddhist agents who carried out the stream of the religion of Buddhism. That is enigmatic indeed since it shows that historical agents at the level of 'enlightenment' still are unable to fully free themselves from the historical determination of the eonic sequence. There is some 'causality of freedom' we don't see since the so-called 'fourth state' beyond self-consciousness (*turiya*) can emerge, not only in relation to the efforts of individuals, but on schedule in an historical sequence. The sudden appearance of a 'Buddha' on cue in a matrix of periodization seems to contradict assumptions about historical transcendence. There is some higher power we do not grasp behind this, although we see it is connected to evolution in our sense. In any case, the eonic sequence comes out 'clean', untampered with in its scale and prodigious variety by the manipulations of spiritual agents. These figures give themselves away with their preoccupation with 'founders' at t-zero initialization points, and are not in a position to even observe, let alone exert authority over the direction of evolution (transitions of several centuries in length, globally dispersed over millennia), and were clearly unaware of the larger process to which they powerfully contributed.

2.53 Spinoza

> **Islam** It is clear from our model why the Axial religions began to crystallize about two centuries after -600, as the transitions wane. Our list of transitions was minimal and might have included the parallel Zoroastrian tradition that will influence Islam. We have spoken of the eonic emergence of religion, but this is misleading if it is seen as deterministic causal generation from sources. For the steps of construction, although echoing their sources, show little that was predestined. The point should be clear in the fanning process of the several 'islams', with the original Judaic tugboat proceeding on its own way.

But these religions accomplish their missions, in many ways. A foundation is laid for passing beyond slavery, for new types of social existence. That the Judaic tradition proved more capably potential for this task than the Hellenic is a reminder of the efficacy of parallel emergentism with its multiple potentials. The picture is difficult to resolve accurately. Was the post-Exilic Judaism a firebrand revolutionary force moving against the past, or a 'steady as she goes' conservative force maintaining a variant of the ancient Mesopotamian temple tradition in a new upgraded form? In any case, the 'myth of Exodus' expresses beautifully the 'virtual revolution' behind the eonic revolution in a tale, as noted, dated precisely to the generations near the divide, or later. The classical phase shows at its clearest our 'fundamental unit' in action, the creation of a bouquet of multiple oikoumenes, from China to the West, as separate yet intersecting cones of diffusion that fall short of global closure. History has outsmarted the one-track mind, with a hope against the imperialists.

The emergence of a world civilization would seem the achievement of the modern transition. It is arguable that a 'world civilization' was already coming into existence from the period after the Sumerian. Within a few centuries the implications of 'first civilization' were already generating a first world civilization around the Sumerian generator as the expanding field of civilization passed into its Akkadian expansion. Whatever the case, the classical world lays the real foundation for global civilization, even as it spawns its characteristic 'islams' in the occident.

The abrupt appearance of Islam at the exact middle of the great passage of our second cycle is hardly surprising. Like the engagement of a pusher unit on a freight train, to move sluggish tons over a mountain range, the effect of this 'man-made' jump-start was decisive, in many ways, with respect to the chaos of occidental antiquity. The same can be said, to a lesser extent, of medieval Christianity, of which Islam is all too obviously a brilliantly streamlined upgrade, ditching the hopeless metaphysical baggage of this trial-run. The issues in the time of Mohammed were very real. Twelve hundred years of coordinated civilization had fallen to pieces. Men, who could see, were aghast at the situation in which they found themselves, at the climax of cyclical downturn.

That this generation of a whole new religious civilization was 'mideonic free action' rather than phase generation, i.e. no exception to our pattern, can be seen from many clues, preeminent among them the fact that one prophet was able to precipitate a 'butterfly effect' against the disorganization of the times.

2.9 The Rise of the Modern: A Second Axial Age?

Almost as remarkable as the sudden onset of the Axial Age is its sudden waning and the return of what we should almost call 'history as usual'. There is something odd about it. The world against which the Axial phenomenon reacts was itself a kind of middle age. And the succession to the Axial period is another. We are left to wonder what the significance of the Axial Age might be. And most of all we are confronted with a question of dynamics. And we are confronted with something unlikely: the uniqueness of this period. Jaspers' use of the term 'axial' is ambiguous in that respect. It seems to point to a unique period in history, a pivot point. But a larger look at world history suggests something quite different, a succession of 'axial' periods.

The Enigma of the Axial Age
Against the Backdrop of World History

We saw our pattern as a short series of epochs in succession each associated with a kind of punctuated transition at its onset. Since the data becomes clear only with the invention of writing, which occurs in the first of these transitions, the data for the first phase is just on the threshold. The resulting fragment is nonetheless unmistakable in its structured coherence. We can see this pattern from several perspectives:

1. The first, visible from our outline of world history, is of the mysterious drumbeat pattern of epochs in world history, proceeding down a mainline of the diversity of civilizations. Note that these turning points are equally spaced, with an interval of about 2400 years, clear evidence of a cyclical phenomenon.

2. The second, which is really an aspect of the first, is of the so-called Axial Age, the enigmatic synchronous emergence of cultural innovations and advances across Eurasia in the period of the Classical Greeks and early Romans, the Prophets of Israel, the era of the Upanishads and Buddhism in India, and Confucius in China. We could have discovered our pattern from analyzing this period in isolation. Looking at this Axial phenomenon we are forced to consider that it is really a step in a sequence, and moving backwards and forwards we suddenly discover the full pattern.

3. The decline and fall of ancient civilizations followed by the sudden rise of the modern world after 1500 is a puzzle that has long confounded world historians. But the puzzle is easily solved if we extend the domain of analysis to include the whole of world history. The puzzle of modernity falls into place in the larger puzzle.

4. The sudden take off of Sumer and dynastic Egypt in the centuries just before -3000 again suddenly falls into place with a simple explanation, not as the 'beginning' of civilization, but as another kind of 'axial' turning point, such as we see in the subsequent 'Axial Age'. This kind of 'relative beginning' phenomenon, like tree rings in an annual pattern, makes complete sense, but requires getting used to when applied to world history.

5. We are left to wonder if this series has a starting point, perhaps in the Neolithic. It is clear evidence of the existence of a 'driver', thus of directionality, the great taboo. But now the evidence is clear.

The clear traces of a non-random pattern taking the form of a sequential logic is a giveaway to some kind of evolutionary process. Note that this kind of intermittent process answers to the paradoxes listed in the Preface. DMR

A Riddle Resolved 187

We have but to zoom out to see that a very simple pattern is at work in the progression of civilizations since the Neolithic. Jaspers himself attempts to generalize his finding, but is obstructed by the issue of 'civilizations'. And his examination of modernity is on the threshold of discovering a 'second axial age', but is thrown off the scent by the confusions of secularism.

It is odd at first to consider the solution to be a frequency hypothesis, but, whatever the case, the basic facts speak for themselves: the Axial Age is part of a larger sequential structure. We should start moving in two directions, backward toward the Neolithic and forward toward the present. The 'axial' character of modernity is often noticed. Thus Bruce Mazlish observes, "The German philosopher Karl Jaspers has spoken of the periods when the great religions arose as 'axial periods'. At such times, there is a 'revolution' in the conditions of human existence and society turns on its axis."[24]

> **Postmodern riddle explained?** All at once, if we can trust the analogy, we see why the sense of a 'postmodern' age arises: it is not the decline of a civilization, but the waning of an impetus, clearly visible after the Axial interval, that mimics 'decline'. Out postmodern confusion is a similar reaction to the immense impetus of the rise of the modern.

We should begin to backtrack to find the 'axial' before the 'Axial'. Joseph Campbell finds an axial period at the dawn of Sumer. The Sumerian source is easy to underestimate. It looks primitive to us now, but its immediacy of creative surging gives birth to 'real civilization' in the odd 'early hybrid modern' where the village passes to the large city-complex. Its effect must have been as seminal as the later Greek transitional era to those who received its influences. It is as if everything was invented all at once, in embryo, to constitute the root-ideas of coming civilization. Thus,

> In the epoch of the hieratic city-state (3500-2500 B.C.), the basic cultural traits of all the high civilizations that have flourished since (writing, the wheel, the calendar, mathematics, royalty, priest craft, a system of taxation, bookkeeping, etc.) suddenly appear, prehistory ends, and the literate era dawns. The whole city now, and not simply the temple compound, is conceived of as an imitation on earth of the cosmic order, while a highly differentiated, complexly organized society of specialist, comprising priestly, warrior, merchant, and peasant classes, is found governing all its secular as well as specifically religious affairs according to an astronomically inspired mathematical conception of a

---

24 Bruce Mazlish, *The Meaning of Karl Marx* (Oxford: Oxford University Press, 1984), p. 8.

sort of magical consonance uniting in perfect harmony the universe.[25]

We note the obvious similarity of this statement to Jaspers' observation of the later 'Axial' Age. Describing the swift transition from the era of earliest Egypt, Michael Hoffman, in *Predynastic Egypt*, is driven in some puzzlement to adopt the economic take-off idea of the economist W. W. Rostow as a metaphor to account for the sudden change that produces the unification of Upper and Lower Egypt under the Pharaoh Menes:

> The immediate archaeological problem in explaining the cultural identity of Menes and his state is to account for the sudden embarrassment of riches that characterizes the material culture of Egypt between the Late Gerzean (ca. 3300 BC) and Archaic period (ca. 3100-2700 BC) in terms of a sophisticated, multifaceted explanation. Professor Renfrew borrows the term 'take-off point' from the economist Walter Rostow to characterize the rise of civilization and the proliferation of certain types of artifacts. Over the years a number of propensities develop within a social system, which predisposes it to a really major transformation. When that transformation does occur, it is so thorough as to convey the impression of crossing a critical threshold.[26]

Remarkable, to say the least. What about Mesopotamia? In *Prehistoric Europe*, Philip Van Doren Stern wrestles explicitly with the evolution/revolution paradox and observes the sudden jump to the first level of civilization in the first hydraulic world of Mesopotamia as it emerged from its mysterious roots of it in the era of the so-called Ubaid and before:

> Something happened in Sumer during the fifth millennium B.C., when all the rest of the world was still so primitive that the Sumerians had to make their own way. The initial stages proceeded slowly for a thousand years or more, and then, during the five centuries between 3300 and 2800 B.C., culture accelerated so rapidly that in this brief time villages became cities and cities grew into city-states...Roux[Georges Roux, *Ancient Iraq*, London. 1964,] merely says of this extraordinarily rapid cultural development in Sumer that 'a close examination reveals no drastic changes in social organization, no real break in architectural or in religious traditions. We are confronted here, not with sudden revolution, but with the final term of an evolution which had started in Mesopotamia itself several centuries before.' Perhaps. But perhaps he is applying our modern time scale to an age when centuries were equivalent to our decades. For a village to become a city in a few hundred

---

25 Joseph Campbell, *Primitive Mythology, Masks of God*, (New York: Penguin, 1959), p. 404
26 Michael Hoffman, *Predynastic Egypt*, "In Search of Menes".

years when there had never been a city anywhere before, is, to put it mildly, something more than ordinary evolution.[27]

Again, remarkable. And this statement suggests we can keep on going backward to find a still earlier case, but for the moment we have discovered something very simple, and a resolution, to some extent, of the riddle of the Axial Age, it is but one in a series. There is one last piece to our puzzle,

2.54 Peasants War, 1525

the rise of the modern. Having moved backwards toward the beginning of civilization, we can move forward from the Axial period.

The sudden waning of the Axial effect, as we have noted, is dramatic. By -200 the Axial phenomenon is clearly over, and the onset of empire seems like a rush into a vacuum, to replace a brief period of republican experiments. The onset of the Hellenistic world of empire is almost a return to the world whence the Greek experiment hopes to escape. In the case of Greece the period of spectacular achievements is over as the Hellenistic, soon yielding to the Roman world ushers in the age of great empires. It is interesting to consider the cognate relation of the Greeks and the Romans, and to consider that the early appearance of Rome and its republic is really a part of the Greek phenomenon. As we study the Greeks we note the way in which their common culture was a function of language and custom, and that this was in turn a medium binding a set of city states and their colonies across the Mediterranean, including the southern part of Italy.

---

27 Philip Van Doren Stern, *Prehistoric Europe* (New York: Norton, 1969)

Was not Rome, in a sense, a child of that nexus of all things Greek, as the diffusion of ideas and the vague sense of a new age animated those in the immediate field of Hellenic influence?

Thus, the emergence of Republican Rome is really still another branch of our far-flung Axial Age, and the appearance of the Roman Republic is the cousin to the surge of republican experiments in the age of Greek political innovations, and the uniquely prophetic creation of the world's first democracy in Greece. There is something significant in the brevity of the Athenian experiment, and the endurance of the Roman. The Athenians will leave a hope for the future, not to be realized until millennia later, in the rise of the modern world. The Romans will carry the issue in its sturdy republican form until the onset of its imperial phases precipitates finally the breakdown of its phase in Axial swaddling clothes and the age of the Caesars begins, enduring all the way into the medieval period.

There is something odd about our use of the term 'middle ages'. We spontaneously consider that the era after the fall of Rome is the middle of something. In fact, it is in the middle between the Axial Age, as a boundary point, with its associated Roman continuation, and the rise of the modern world millennia later. This 'medieval period' suffers a charge against its reputation in our minds, then, one frequently protested by various parties to its defense, in the way we see it as in some fashion not up to the standard of either its Axial beginning point or its modern recurrence. Whether this downplaying of the medieval interval is fair or not, the fact remains that our very terminology reflects a larger pattern of history, and on a scale that goes far toward explaining why a pattern of overall coherence is hard for us to detect. For until the rise of modern archaeology the beginnings of our traditions seemed to be those visible in the Axial period. The intimations of unknown earlier acts of the play are seen in the unexplained appearance in Biblical history of the Egyptians, or Assyrians, lurking in the background as remnants of some unknown world thought to be passing away.

This effect of relative beginning in what we have dubbed the 'Axial Age' seems then to suggest a complete unit, of 'punctuation' and the 'equilibrium' that follows in its middle period, until what is apparently another punctuation occurs, and this we call the rise of the modern world. We are getting suspicious. If the Axial Age is a kind of new beginning inside a larger history, its uniqueness would seem to have been the result of our lack of knowledge of earlier civilizations. But this lack of knowledge about the earlier stages of civilization is no longer the case: the rise of archaeology has shown us the antecedents for the mysterious Assyrians and Egyptians who appear in the Biblical text. And as we proceed backwards we are left to wonder if some antecedent 'Axial' period is not visible in the historical image crystallizing in archaeological fixer. We already know the answer, if indeed we are aware of any of the findings of modern archaeology, which show us the so-called rise of civilization at the end of the fourth millennium in strangely synchronous emergence of Egyptian and Sumerian civilizations. Strange to say, we can even produce a rough interval between these moments, of just over two millennia.

The dynamism of the Axial period, its seminal creativity, seems to fret an entire an entire cycle of civilizations, and is unmatched by anything until the rise of the modern world. What is remarkable is the loss of so many of the innovations of the Axial period, a notable example being the birth of science, and its slow passing away with time, such that by time of the medieval period, in the Christian West, its birth among the Greeks is almost a forgotten memory. Its partial survival in the world of Islam is like an ember fire carried across time.

And then suddenly in the sixteenth century we see once again, almost like a timed renewal, what is in many ways a recursion of many of the innovations of the Axial period, with some important differences. The parallel transformations of the Protestant Reformation and the Scientific Revolution, Copernicus and Luther, stand at the threshold of the modern transformation leading to the rough point, around 1800, when a transition to a new era seems complete, and a new age begins, at the threshold of globalization. The phenomenon of the rise of modernity is the object of many theories and controversies, but the basic observations of the phenomenon resemble the exclamations we find with the Axial Age.

There is a mysterious seminal generation springing from the period ca. 1500, indicated by the onset of the Reformation. Over and over our sense of historical modernism draws us to this point of the so-called 'early modern', and into a controversy or equivocation over its significance as one of the great turning points of history. Relative to world history, progress explodes in the sixteenth century, despite the puzzle over the Renaissance. The abrupt start after 1500 is constantly suggested and then challenged or retracted because its proponents cannot account for it, or sort out the fact that a discontinuity might interrupt prior continuity.

This sudden change in direction is reflected in the puzzled observations of a host of historians. J. M. Roberts in his *History of the World* opens by noting, "After 1500 or so, there are many signs that a new age of world history is beginning…". William MacNeill, in his *The Rise of the West*, calls the career of Western civilization since 1500 a vast explosion. Geoffrey Barraclough, in *Turning Points in World History*, notes the remark of Paul Valery that Europe is a 'peninsula of Asia', a western appendix of the Eurasian land mass, and asks, "How was it that this western appendix came to be in a position to exercise this power, this domination over the greater part of the world?" He cites the factors of technological and scientific proficiency, the revolution in transport and communications, that 'caused' this brief hegemony, but in a manner typical of historians stumbling over the eonic effect is driven to note, "So much, I think, is obvious; but it tells us very little".[28]

Fig. 2.55 *Mayflower* compact

---

28 J. M. Roberts, *The Penguin History of the World* (New York: Penguin, 1990), p. 526. Cf. also, p. 529, for a discussion of the relativity of the term 'modern', which was once inclusive of the medieval, then distinguished from it, and now might be distinguished from the contemporary by a new term, the 'early modern'. L. S. Stavrianos, in *The World Since 1500* (Englewood Cliffs, New Jersey: Prentice-Hall, 1975), "Why should world history begin with the year 1500?"
It is significant the term 'medieval' was itself a child of this period, or that just after, when the German scholar Kellarius coined the term '*Medium Aevum*' to distinguish the suddenly apparent new 'modernity' from the 'middle period' after the fall of the Roman Empire. This fact is another caution to those who use the term 'Renaissance', a concept created in the nineteenth century. Men of the sixteenth century did not use it, but were stunned by the sudden changes before them, as they expressed, not a rebirth, but the rise to an entirely new form of complex civilization.
William MacNeill, *The Rise of the West* (Chicago: University of Chicago Press, 1963), p. 567. William A. Green, *History, Historians, and the Dynamics of Change* (Westport: Praeger, 1993. Jacques Barzun, *From Dawn to Decadence*, New York: HarperCollins, 2000, p. xvii. Geoffrey Barraclough, *Turning points in World History* (Great Britain: Thames and Hudson, 1979), p. 3.

A Riddle Resolved

Marshall Hodgson, in *The Venture of Islam*, speaks of the Western Transmutation, 1600 to 1800, and sees the connection with the earlier period, generated from Sumer, but his analysis focuses on the history of technology, and fast-forwards to exclude the Reformation.

Fig. 2.56 Luther tears up papal bulls

What happened can be compared with the first advent several thousand years BC of that combination, among the dominant elements of certain societies, of urban living, literacy, and generally complex social and cultural organization, which we call civilization.[29]

Jacques Barzun in *From Dawn to Decadence* asks, "Granted for the sake of argument that 'our culture' may be ending, why the slice of 500 years [from 1500 to the present]? What makes it a unity? The starting date 1500 follows usage: textbooks from time immemorial have called it the beginning of the Modern Era." There is no implication of decline or decadence after the interval of transition, since a new era has come into being. The conclusion of the eonic sequence should be great new beginning.[30]

This sudden take-off (relative to world history) has always been

---

29 Marshall Hodgson, *The Venture of Islam*, Chicago: Chicago University Press, 1974, 179. See also, *Rethinking World History* (Cambridge: Cambridge University Press, 1993), Marshall Hodgson, Edmund Burke III (ed.) (1993), Ch. 4, "The Great Western Transmutation".
30 Jacques Barzun, *From Dawn To Decadence* (New York: HarperCollins, 2000), p. xvii.

intractable for students of the question, and driven historical sociology into a frenzy of Renaissance resurrections, dialectical Big Bumps, Marxist social stages, Weberian econo-religious explanations, or the 'European Miracle' of the historian E. L. Jones.[31]

Figs. 57, 8 Thomas More, *Utopia*

As noted, the periodization question of the 'rise of modern' has many casualties in the realm of theories. Three sets of failed theories deal with these eras in isolation, those of the rise of the modern, the birth of civilization, and, to the extent they exist at all, efforts to explain the Axial period, along with the whole spectrum of interpretations of the classical civilizations, to say nothing of explaining the history indicated in the Old Testament. Without exception these theories have all failed. Suddenly we realize they are really all asking a similar set of questions about an invariant puzzle. The question of the 'modern' remains baffling until we see it in its greater context. Then the remarkable resemblance of the rise of the modern to the Axial interval, and especially Greek Archaic appears.

We are closing in on a pattern of universal history, at once simple, and mysterious, and clearly showing us the principle of coherence we were seeking in our perception of world history. And we are close to the resolution of the riddle of modernity, and to a perspective on the way it might suddenly show chaotification. We seem to be, not in the stages of the postmodern, but in the early stages of a great new era of world history, after passing through the

---

31 E. L. Jones, *The European Miracle* (New York: Cambridge University Press, 1961).

transitional period of its onset. And as we explore this larger framework we can attempt to redefine the modern in a fashion more conducive to the needs of our future, beyond the domination of economic fundamentalism, or the imposition of false views of evolution on the outcome of something larger than Social Darwinist paranoia and environmental degradation. We begin to see the clue to better resolution than the return to traditionalism.

> **Democratic Revolutions** One of the most mysterious aspects of our new perspective is the double birth of democracy, in classical Greece and the modern transition. This exact correlation is one of the most remarkable discoveries of careful periodization, and leaves us to wonder what it means.

Fig. 59 Thomas Münzer

As we examine this 'ratchet effect', the pattern confuses us because it does not follow the course of a single civilization, but jumps between civilizations as it proceeds. The question of the rise of the modern world also shows the displacement of change beyond the frontiers of the old Roman Empire into those parts of Europe that were only marginally a part of the ancient Roman system. We observe the Reformation, and see a religious phenomenon, but we might look beyond religion to see the opening of a new field of culture free from and at the exterior to the system of antiquity. In fact, we begin to sense another instance of the frontier phenomenon that we noted in the Greek Axial Age. This is in many ways the signature of this age of renewal, as it expands beyond the framework of antiquity, first to Northern Europe, thence to the Americas, and beyond. We must begin to wonder if the phenomenon we are trying to understand is not a globalization process more than a phenomenon of civilizations.

Our sense of modernity has been confounded by a false Eurocentrism, but we can begin to see beyond that. The constant references to 'Western Civilization', or the 'West', or the Judaeo-Christian heritage, in a series of Eurocentric terms, blinds us to the reality, which is that the rise of the modern is not a European phenomenon, as such, and finds its field of realization almost sooner in its exterior than in its homeland. The obvious picture left by history here is the temporal correlation of the spread of European, we should rather say, Eurasian, civilization to the Americas. It is hardly accidental that the North American colonies beginning in the seventeenth century already show the seeds sown by the English Civil War that will grow later in the classic harbinger of a new era dawning, the American Revolution.

There is obvious something larger than Europe then in the modern transformation and the result is the birth as much of a new global civilization as the passage of a cultural particularity called the European. The same interval of sudden change, followed by the creation of an oikoumene in the diffusion from a source, is visible in the modern world as it was in the Axial Age of Greeks. And a comparison of the two leaves us with a set of unanswered questions about the nature of historical change, and the more general issue of slow or fast evolution. We seem to see, or think we see, the slow evolution of modernity from a medieval world. But it resembles very closely the Greek Axial interval, and there we were left hanging with such explanations. There wasn't anything at all slow about the Greek Miracle. In a few centuries it emerged from nothing, flowered in spectacular fashion, and was done. The sense of a resemblance with the modern transformation begins to suggest a new and different kind of explanation for the rise of the world we have inherited from the early moderns.

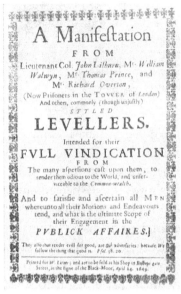

2.60 *Manifesto of the Levellers*

2.9.1 A New Age Begins

'We are at the dawn of a new era!' exclaimed Luther more prophetically than he himself imagined... 'Rarely is a work undertaken out of wisdom and precaution,' he declared, 'but everything is undertaken out of ignorance.' The man who initiates creative action can seldom know where his steps will lead him...But if Luther was a prime mover, the forces that soon set all Europe in motion were stronger than any single man.' "

Returning full circle from our search for the sources of the eonic sequence we arrive once again at the dawn of modernity to find our world system taking off on schedule in the sixteenth century in one of the last diffusion frontiers left, spawning the new era that we call modernity. The rise of modern is now transparent as the third great transition in our eonic sequence. We are back at our starting point with a structure of elegant, yet mysterious, coherence that highlights two different levels at work in world history. Despite these theoretical-sounding statements, the pattern of the eonic effect, let us remind our selves, is purely empirical, however we understand it, and the sudden rise of the modern world from the sixteenth century onward is a mysterious given of world history, and completes the equal mystery of the previous two transitions we have examined.

All the confusions of discontinuity, Eurocentrism, and secularism, disappear in the expanded scale of our eonic analysis. The rise of the modern is not a development of a Western Civilization, but an eonic transition expressing world-historical directionality of a cluster of culture complexes in a frontier effect: North Italy, Spain, France, the Protestant Crescent (Germany, Holland, England, and soon, its sidewinder, North America). This transitional phase is over by the end of the Enlightenment, and the system rapidly starts to globalize on this new basis, in the slow shift of the center of gravity. Once again our eonic sequence hazards its globalization on a temporary localization and the immense strain of macro-action via micro-action soon finds democratic emergentism competing with imperialism and revolution. We should note that globalization in our sense is a function of the eonic sequence, and not the same as economic globalization.

As if the last place left on the planet to stage a surprise attack against Eurasian inertia the Euro-partition created by the Reformation generates a new frontier sector that takes off in a race against time and newly expanding slavery, in the brief launch window, closing if not closed, by the rough point of the divide, before the underdog becomes a new source of domination and empire. Democracy comes roaring back, much stronger this time, abolition is achieved, and it almost seems as if the Ionian Enlightenment is in a second coming against the theocratic worlds created by the winners of the Axial period. We can add the 'rise of the modern', now a time-slice phase, to our list of stream and sequence intersections, resetting the directionality of the world system as it moves toward globalization.

We can see how this transition forms a coherent unit in two rough halves as the Reformation and the Copernican Revolution leading past the Thirty Years War brings us to the new age of the Enlightenment, renewed democracy, and the Industrial Revolution. Although past the modern divide, we are still altogether in the grips of the modern transition, and culture still has the freshness of a new age in world history, despite the convulsions of the past two centuries and the onset of postmodern chaotification in the waning of the elusive factor of eonic determination.

The resemblance to the Greek transition is striking, almost like a recursion. The immense potential lost in the post-Axial chaotification of the Hellenistic seems to get a second chance. Let us note that science, including the idea of evolution, and democracy both failed the 'survival of the fittest' test, the case of the missing centuries, and show our clear evidence of eonic mainline reinduction. So much for Darwinian thinking. Our univalent modern transition, compared to the Axial parallelism, is severely imbalanced in one sense, leading to Eurocentric illusions, but the overall logic is clear, and the swift turn toward cultural globalization occurs promptly in the wake of the divide, thwarted by the forces of rising imperialism.

The phenomenon of Axial parallelism would be counterproductive in the modern transition, and the emergence of universalist themes is a striking feature of the Enlightenment contribution to globalization, real globalization. Alone among the great religions the Christian stream is in the eonic mainline and the swift remorphing of its Protestant trigger into the Enlightenment shows the deft effectiveness of the transitional era. Our model renders no judgment as to either the true definition of religion, or its future in the world system. In one sense, as secularists would believe, religion is a redundant category, from the view of our fundamental unit of historical analysis. But it would be naïve in the extreme to pronounce on the future passing of religion, as the host of New Age movements, to say nothing of the leftist themes of class struggle, already show the trend toward mideonic reformulation of religious fundamentals. The issue is not religion, as such, but the inability of all parties to create spiritual vehicles that are not vehicles of exploitation, or domination.

It is thus significant that many now sense what they call a 'postmodern' age. Our interpretation shows the reason, and the paradox of progress surging, progress in paradox. This term is superfluous in our model and postmodernist periodization tends to create confusion, whatever our views on its philosophies, where a 'dialectic of the Enlightenment' is simply par for the course. As a critique of teleological ideologies postmodernist thinking is

A Riddle Resolved

2.61 Leibnitz

significant. But we might just as well critique a lack of a true universal history, equally able to produce a 'postmodern' assessment of our historical dynamics.

Our interpretation deftly bypasses the illusions of Eurocentrism and we see that the eonic sequence is moving on a far greater scale than that of individual civilizations, if only it can become disentangled from the local medium of its long-range action. Our system can generate change in the core, but cannot control its peripheries, the undoubted reason such an explosive left arose so quickly in the wake of our transition to challenge the instant distortions of globalization. Our modern transition is not the triumph of 'Western Civilization' but a pivot on the way toward globalization. And this globalization is not the same as economic development. That is true by definition in our account, but clearly economic action rapidly becomes the key player in this instance. If we compare the three centuries of the ancient Axial transitions, plus the two centuries immediately in their wake, then look at the modern instance, as five centuries from the onset of modernity, we see it is not surprising and no accident to find the current preoccupation with empire and pseudo-globalization of economic exploitation. It is almost too mechanically precise for comfort.

2.62 Leibnitz invented differentials and Newton fluxions

Well past our divide period, the world system is now in the throes of its reversal toward the whole, and our model is ready with its balance of two universal histories in the dialectic of universalism and diversity. Chauvinist or Eurocentric accounts of our modern transition (e.g. the 'Judeo-Christian tradition, etc,…) will be swiftly disabused of their sense of centrality as the system slowly but surely changes its center of gravity. In fact, the first shift in that center of gravity occurred early on in the American sidewinder. The latter would do well to consider the gifts of time, not overestimate one's brilliance, and not fall behind as the globalization process continues. We should not forget that, while our use of the term 'evolution' is at risk of an ethnocentrism reflecting the transition zones, its scope in reality is universal, and moves to garland the fruits not only of its prior stages, but of the universal dimension of evolution in the greater community of man irregardless of its coordinates in relation to the eonic sequence.

By our analysis, instead of a postmodern, we are in a post-transitional period, a better way to put it, still close to onset of a great New Age of world history, whose potential we must hope will not end betrayed as have prior stages of civilization. If postmodern philosophies echo and descant the very Enlightenment they critique, then they join that canon in reasonable fashion. But if the idea is to replace the modern transition with a new New Age negating the rise of the modern, the odds against success are very great, unless simple decline is a possible candidate. Although in a postmodern period the rise of the modern and the Enlightenment are under attack and the critique of imperialism and empire seems to replace the discourse of democracy, our emphasis on the early modern is the right one, in terms of the overall 'eonic evolution of civilization'.

2.63 Kant

Our transition is taken as the dawn of a New Age. The mythology of New Ages is unending, but our eonic mainline gives us a useful way to set the record straight and we can categorize the modern transition as the dawn of a New Age in some hope to still the commotion here. Although our use of the idea of a 'New Age' is informal, and has no theoretical status, we can, for all intents and purposes, depict the third transition as rapidly emerging modernism in terms of a 'New Age', the third in visible world history, the more so as its challenge to the outstanding religions of antiquity is so reminiscent of the 'relative transformations' of the Axial period. Beware of those pronouncing the Enlightenment a failure and proclaiming the new New Age for some guru or others ambitious to exploit a postmodern strategy.

We have almost whimsically taken on the lore of cyclical theories, to challenge the Spenglers and Toynbees. Our data shows the correct grounds for this, but does not allow us any empirical generalization. So we merely observe the factual mystery of a cyclical phenomenon first visible in the era of early Egypt and Sumer. We must be clear we are speaking of

A Riddle Resolved

cyclical progression, empirically given as with economic cycles, and not cyclical recurrence in some metaphysical phantasm of cycles. The cyclical progression of 'Mondays' in a sequence of weeks is not the same as the cyclical recurrence of their interior events. One reason to produce a 'cyclical' theory at all is to challenge the prophets of doom and decline who will attempt to point to some 'decline of the West' as a postmodern comeback against modernity. This view reconciles perfectly the 'opposed' linear and cyclical views of history and gives new meaning to ideas of evolutionary progress. Our viewpoint reconciles the so-called linear and cyclical views of history into one concept.

Fig. 2.64 Adam Smith

The center of gravity of our modern post-transition might well change, but this is not an issue of the imperial powers of the first and early inheritors of the modern system. It is good to be wary of the Toynbean formulation. Toynbee begrudged the modern world the breakthrough Enlightenment, and seems to find at the point of globalization the need for religion as some phantom of the internal proletariat. We are wise to this game. These religions are mostly mideonic sludge at this point, and don't correspond to the Axial source.

2.9.2 *From Reformation to Revolution*

Of all of our transitions, the modern is the most transparent because we have continuous data throughout, and the result shows a clear overall dynamic and interior structure, in a unity stretching from the Reformation and Copernican Revolution to the Enlightenment and French/American Revolutions. And this transition falls naturally into two stages, centered on the seventeenth century, as the Reformation ignites the fast passage, the field clearing in the wake of the Thirty Years War, to give birth to the seminal first signs of virtually all the characteristic eonic emergents of modernity. The relative transformation of a small piece of Christendom on a northern frontier, the Protestant Reformation, is a classic instance of the 'eonic evolution of religion'. This 're-formation' is at first confusing in that

Fig. 65 Schopenhauer

This remarkable new model of yogic discussions of consciousness is a de facto 'Reformation' of Buddhism for modern times!

A yogi in a new guise In the wake of Kant the philosopher Schopenhauer produced a brilliant, streamlined version of transcendental idealism. We might cite a passage from Dale Jacquette's *The Philosophy of Schopenhauer*, remarkable for revealing the latent potential of 'transcendental idealism'.

Schopenhauer's philosophy often gives the impression of having been composed expressly for the purpose of reconciling the phenomenal will to the inevitability of death. All the apparatus of his main treatise, the fundamental distinction between the world as Will and representation, the concept of thing-in-itself as beyond the *principium individuationis*, and fourfold root of the principle of sufficient reason, can be understood as contributing to a moral, metaphysical and mystical religious recognition that death is nothing real and hence nothing to fear. If Schopenhauer is correct, he proves that death is not an event, and hence altogether unreal. Death is not an event in the world as representation, but is rather an endpoint or limit of the world as representation, and in particular in the first-person formulation as my representation. The world as representation begins and ends with the consciousness of the individual representing subject. At the moment of death, all representation comes to an immediate abrupt end, after which there remains only thing-in-itself. An individual's death is not something that occurs in or as any part of the world as representation. Nor can death possibly be in or a part of the world as thing-in-itself or Will. There are no events or individuated occurrences, nothing happening in space or time, for thing-in-itself, and in particular there is no progressive transition from life to death or from consciousness to unconsciousness. If with Schopenhauer we assume that there exists only the world as representation and as thing-in-itself interpreted as Will, then there is no place on either side of the great divide for death, no possibility for the existence or reality of death.

it is a religious rebirth that remorphs into secularism.

The early modern: an emergent field Let's list a few of the eonic emergents relevant to our definition of the modern *transition*. Although the size of this dataset is staggering, if we list enough overlapping zoom targets we can likely get a fair picture of what's going on. The list can keep growing. We are *outside* this transition, and must assess using *judgment* what should be on the list. But even with a partial or debatable list we can make our point, TP3 creates a massive change of historical direction. Thus we get:

The Reformation, with Luther's and Tyndale's Bible, Copernicus, Vesalius, then the seventeenth century Scientific Revolution, the birth of liberalism, Descartes and the rise of modern philosophy, Hobbes and onward, the German, English, American and French Revolutions, the birth of democracy, the Enlightenment. The Industrial Revolution, and the onset of modern capitalism…

2.66 Monteverdi The spectacular version of the 'frontier effect' in the modern transition, the northern crescent includes northern Italy which is the source of the stunning fast development and flowering of modern music climaxing near the 'divide'. Monteverdi is a good example of the stream and sequence dynamic as he starts the great music sequence out of Renaissance sources.

Note that the generation near the American Revolution, our divide inside our transition, is one of the most massively packed periods of innovation in world history, and much more than a matter of technical innovations.

We see the French and American Revolutions (and soon liberalism spawning democratic liberalism), the Industrial Revolution, the Enlightenment with a Scottish Enlightenment, and a German *Aufklärung*, Adam Smith and a new economics, German Classical philosophy and the Romantic Movement, Kant, Hume, Bentham, Thomas Paine, … This just skims the most visible data off the top. Our divide is a matter of degree, and could be from 1750 to 1850. But there is a clear fall off in the rate of *basic cultural* innovations, as opposed to technical innovations or economic expansions. A good way to see that is in the Industrial Revolution. That creates a massive transition of its own, and then stabilizes as a 'market society', however unstable that is.

TP3+: Since our turning point is a finite interval, it produces a divide (early nineteenth century?) and, sooner or later, goes through a post-transitional phase, perhaps of reaction against the turning point.

Fig. 67, 68 Hegel and Marx

The onset of the modern transition shows us a mysterious starting chord in the synchronous appearance of Luther, and Münzer, next to Machiavelli and our first modern Utopian Thomas More. Let us remind ourselves that if Machiavelli initiates a new science of politics, the hidden note of politically invisible actors, no doubt immoral riff-raff, mongrel descendants of the godly Pharaohs, it is also true that precisely at our divide an ultra-idealistic protest, anti-Machiavel, appears in the Kantian contretemps with Benjamin Constant. Before continuing we should rescue our subject for some 'idealistic thinking' with an interpolated 'sermon' in the midst of 'value free science'. Realist politics and the devious schemes of Machiavelli have no status in our system.

The Northern Crescent In relation to the frontier effect, the prime transitional zones lie along a Northern Crescent, with an early trigger in Northern Italy: Germany, Holland, England, France, Spain. The North American sidewinder rapidly initializes and by the divide point is a prime emergence zone. Our transition has to risk Eurocentrism, then start a fast getaway after the divide: globalization via localization. We are not talking about Western Civilization, or Europe.

Luther—and Münzer Luther's 'revolution' is a geopolitical one, the decisive stroke against the theocratic empire of Christendom, and his 're-formation' is the classic instance of the 'relative transform' effect, so characteristic of our eonic sequence: break off a piece of the prior state of affairs, and remorph that in a frontier effect. Neglected in the overall portrait is the German social revolution of 1625, and the appearance of the first of our radical eschatological champions of the

A Riddle Resolved

The work of Marx and Engels in the post-Hegelian generation up to 1848 is an absolute *tour de force*, but also a confusing history. The tale is told in many works, such as *From Hegel to Marx*, Hook, 1934. Marx's *Economic and Philosophic Manuscripts of 1844* of this period have resurfaced...The larger framework of the Kantian revolution, and beyond that the Protestant Reformation, is essential background.

> All of which is only another way of saying that...it is our affair... to participate in the redemption by laying aside our immediate subjectivity (putting off the old Adam), and getting to know God as our true and essential self...
>
> Hegel

The Self As God
In German Philosophy

The movement of thought from Kant to Hegel revolved in a fundamental sense around the idea of man's self-realization as a god-like being, or alternatively as God...

Robert Tucker, *Philosophy and Myth in Karl Marx* (Cambridge, 1964)

Many books on marxism behind the veneer of Cold War cavil actually have useful insights into Marx, as here with Tucker's attack which backfires and exposes the richness of Marx' thinking and that of the generation post Hegel to 1848

The number of critiques of Marx is large and it can be difficult to sort out the counterrevolutionary unconscious in the philosophic assaults on the legacy of 1848. In *The German Ideology* Marx and Engels are thoroughly mid-fray. Robert Tucker's useful book is also indicative of the shallowness of much religious thought in the West, even as it raises a number of key points, The idea of 'self-realization' is as old as the Upanishads, and to castigate it in the fashion of Christian theism misses the point that millennia of yogis have come to 'know god as the essential self', almost a cliche of sutric discourse. By contrast Westerners self-destruct in a Mephistophelean melodrama. The damnation scenario of Faustian hubris is altogether characteristic of the narrowness of the lesser Christian canon, which figures such as Kant and Hegel attempt to reinterpret. Kant's distinction of noumenon and phenomenon as aspects of the self likewise goes down in flames, but is as an almost primordial rediscovery of Upanishadic *atman/brahman* cloaked in the confusions of his ethical theories. But, as Tucker miscasts him, perfectly describes the 'desert of the ego 'in 'ethical' compulsive neurosis. LFM

proletariat, Münzer.

Machiavelli is often said to initiate the modern era of politics, but he

Fig. 2.69 Hobbes, Locke, Rousseau

is a perfectly Janus-faced figure, looking backward and forward at the same time. As our eonic system starts uphill on Mount Improbable, the world of the Borgias, and the anemic 'renaissance', are left behind, and the counsel to the Prince ends in ambiguity. Machiavellianism has no real status as an 'eonic emergent' except as a token of post-Christianity, but becomes a *de facto* pseudo-standard. But his classic reflections on republicanism will resurface in timely echo at the onset of the American Revolution and the complexity of the integration of separate components of that great new beginning of democracy, or republicanism, both echoes and transcends any interpretation of horizontal politics. Observe how Machiavellian *real politik* is outsmarted by the end of the transition as it touches the ideal, even as the politicians reclaim control of state systems, having learned nothing, but mouthing a different set of slogans.[32]

More's Utopia One translator of More's classic remarks that its position is like that of the baby of the Judgment of Solomon, Catholic tract or political manifesto? It is a premonition, at the least, of the last question spawned by our transition, gestating liberal worlds, the question of private property. In the relative transform of a genre created in antiquity, it spawns the 'eonic emergence' of the utopian genre, perhaps even the genre of science fiction. We should note that our eonic sequence deals in potentials, and utopianism is an exploration of potentiality in relation to horizontally causal history.[33]

---

32 J. G. A. Pocock, *The Machiavellian Moment* (Princeton: Princeton University Press, 1975).
33 Thomas More, *Utopia*, trans. Paul Turner (New York: Penguin, 1965).

A Riddle Resolved

Copernicus The 'eonic evolution of science' in the form of a second Scientific Revolution, the Greek being the first, is a sixteenth century phenomenon, and the 'great paradigm shift' of the Copernican Revolution heralds the first order of business for our eonic sequence, the rebirth of Archimedean physics.

2.70 Descartes

As we examine the modern transition, a puzzle resolved about the Greek Axial interval comes to light: why is the effect of the Greek transition so clustered *after* its divide, and why does the first half of the interval, in the Greek Dark Age, seem to be empty or invisible? In fact, we see the answer in the modern instance. The first half of our transition is hard to distinguish from the 'Middle Ages'. The real onset of 'modernity' occurs in the seventeenth century after the closing of the Thirty Years War. The Greek Reformation, and the progression from monarchy, is there, if we care to look (eschewing overly precise analogies). The first visible effects of the Greek transition appear in the second half, in the eighth century BCE, visible in the Homeric starting point. In a strangely similar pattern, the modern transition really takes off in the generation after Shakespeare and Cervantes, with his Don Quijote, quite the modernist *malgré lui*.

Thus the Treaty of Westphalia tokens the clearing of the field as the seminal gestation of the Enlightenment begins with rise of modern science, philosophy, and the intimations of democracy. We see in the title of the great work by Copernicus, *De Orbis Revolutionibus,* that ushered in the Scientific Revolution both the unfolding, and a new signature definition, of the term 'revolution of the ages', with the ironic new modern meaning for the term, emerging in relation to the other.

The English Civil War The key to the politics of the coming new age is seen in the English Civil War. As Christopher Hill notes in *The Century of Revolution, 1603-1714*, "During the seventeenth century modern English society and a modern state began to take shape, and England's position in the world was transformed", and yet the transformation lies beyond the question of states, the German field having been almost torn to pieces, yet still exhibiting all the elements, by its end, of the transition. The German *Aufklärung* proceeds with or without a state. The seeds of the English exemplar will resurface in the American sidewinder in the emergence of the first great mass

Fig. 2.71 A reading of Voltaire

democracy—at the divide. Christopher Hill, in his *The English Bible and the Seventeenth-century Revolution*, notes the frequent observation that the English Revolution had no 'ideological forebears', that no one passing through it "knew they were living through a revolution", often taking their cue from the Bible![34]

Levellers and True Levellers The period of the English Civil War suddenly spawns a virtual hotbed of diverse and beautifully potential radical movements, from the Levellers to the Diggers and Ranters, prophetic in their import, and leaving behind a legacy that will resurface in the great moment of equalization that emerges at the divide. These virtual eonic emergents that soon disappear remind us that we can never finally conclude the outcomes of our transitions correspond fully to 'what was intended', so to speak. It wasn't long before the same old elites reestablish control. The American Revolution will receive many of the influences appearing at this brief moment of historical

---

34 Christopher Hill, *The English Bible and the Seventeenth-Century Revolution* (New York: Penguin, 1993), p. 7-8. Christopher Hill, *The Century of Revolution, 1603-1714* (New York: Norton, 1961), p. 1.

self-consciousness.[35]

Leviathan: Hobbes to Locke The first seventeenth plateau of the transition produces a recursion from beginnings of political science, with the brutal clarity of Hobbes' opening note, followed by the essence of the future liberalism crystallizing in Locke.[36]

Birth of the Enlightenment The real beginning of the Enlightenment occurs in the seventeenth century with Descartes and Spinoza, and a host of other seminal premonitions of modernity…[37]

The New Atlantis Our transition is not without prophets, in the true 'eonic' sense, and Francis Bacon, although now beset with the critiques of his enthusiasm, creates the ethos of innovation and technological liberation.[38]

2.72 Lagrangain mechanics

The eonic evolution of science Our rubric the 'eonic evolution of X' comes into its own as we observe the nicely scheduled re-ignition of science seen in the (second) Scientific Revolution in our eonic mainline. We should declare the case of the missing centuries solved in noting that the emergence of science is bound up in the 'eonic determination' of the eonic sequence. This raises the question of the contrasting 'science as free action' in the passage to the post-transition. Indeed the crystallization of 'scientism' shows just this effect.

The rise of a distinctly modern philosophy crystallizes with Descartes. As Bryan Magee notes in an account of Schopenhauer, the rise of modern philosophy shows a clear narrative that chaotifies after the period of Kant.[39]

---

35 Christopher Hill, *The World Turned Upside Down* (New York: Penguin, 1991).
36 Craig Thomas, *From Here To There* (New York: HarperPerennial, 1991), Peter Schouls, *Reasoned Freedom: John Locke and Enlightenment* (Ithaca: Cornell University Press, 1992).
37 Jonathan Israel, *Radical Enlightenment: Philosophy and the Making of Modernity 1650-1750* (Oxford: Oxford University Press, 2002), Paul Hazard, *The European Mind* (New York: Penguin, 1964).
38 Charles Witney, *Francis Bacon and Modernity* (New Haven: Yale University Press, 1986).
39 "I have shown how, when the mainstream of modern philosophy ran up against transcendental idealism it ceased to flow along a single current and ramified into various channels." Bryan Magee, *The Philosophy of Schopenhauer* (New York: Clarendon, 1997), p. 96.

Descartes to Hume/Kant The course of Cartesian dualism haunts modernity from beginning to end, and yet if we feel the urge to the non-dual we should consider the plight of contemporary neuroscience shorn of dualistic 'crudities'. Descartes did his work well, and describes the two-sided creature that will inherit the wasteland of Aristotle and Aquinas.[40]

2.73 Mozart

Spinoza It would be hard to find two more 'eonic' beings than Descartes and Spinoza. Spinoza, as if in the first order of business for modernity, appears like an apparition in the Dutch Enlightenment, and produces the last Biblical apochrypha in his brilliant 'exodus', the invention of Biblical Criticism, pantheism, and the foundations of liberal secularism. His thinking proceeds underground then resurfaces at the Great Divide in the famous Pantheism debate.

Perhaps the true resolution is glimpsed at the threshold of awareness, as in Kant's transcendental deduction:

> The Rationalist Descartes takes the 'I think' to indicate the existence of a substance, distinct from the body. This ignores the important paradox concerning consciousness—which is that we cannot experience it, because it is experience. Hence, the saying "the I which sees itself cannot see itself". Kant recognizes this paradoxical point and explains it. According to him, the 'I' is not an object of possible experience, because it is a presupposition of experience.[41]

No Age of Revelation here. All you get is a 'transcendental deduction'. The course of modern philosophy is reflected in this statement, in the endgame of Heidegger, and the postmoderns. As the modern transition takes off into its scientific fugue, Descartes produces a brilliant 'fix' or failsafe that will allow the work to be done by those destined to be left orphaned by the onset of reductionism and its myths, almost as pernicious in potential as those of fanatic monotheism. The work of Kant, and his descanting Schopenhauer, perfectly timed at the divide, will lift the question into a realm evocative of

---

40 William Bluhm, *Force or Freedom?* (New Haven: Yale University Press, 1984), Jerrold Seigel, *The Idea Of The Self* (New York: Cambridge University Press, 2005).
41 Garrett Thomson, *On Kant* (Belmont, Ca.: Wadsworth, 2000).

A Riddle Resolved

the Upansishads, as our eonic sequence comes full circle.

The New Physics The great glory of the modern transition is the birth of the New Physics, with the calculus of Newton and Leibnitz. But the monofocus on the majestic emergence of the new science distracts us from the more complex dynamics and interplay of ideas generated in our transition.

Fig. 2.74 John Keats English (and French) destined as 'koine' for the global oikoumene flood with literary from the start of the modern transition, and Shakespeare, Racine carrying the tragedy enigma. The sudden double whammy of English romantic poets just as the divide is another beautiful effect.

The eighteenth century stages the classic second phase of the Enlightenment and this ends in the rushing cascade of the point of the Great Divide, the generation of revolutions and the emergence of capitalism. This period is massively packed with innovations in all areas and consists of multiple 'enlightenments', the French, English, German, Dutch, Scottish, American,…

Battle of The Ancients and Moderns The classic debate over modernity is the morning songbird of the birth of a new idea of progress, and the passage beyond the achievements of the ancients.[42]

Voltaire, Diderot, D'Holbach Voltaire and the *philosophes* are the spearhead for the secularization process inexorably springing from the Reformation. Diderot with his *Encyclopedie* tokens the 'information revolutions' to come. We should note that Voltaire was not an atheist. The rise of modern of atheism is 'still another eonic emergent', a long suppressed dialectical potential, no more, no less.

Rousseau and Kant Rousseau is in many ways a difficult figure to understand, in part because we think in terms of results, not in terms of the creative dialectical moments of true innovators. Rousseau precipitates the reaction to Newtonianism, the democratic revolution in the evolutionary macro-action of equality/equalization, and is a direct influence on the Kantian analysis of the idea of freedom in the context of the New Physics.

---

[42] Joseph Levine, *The Battle of the Books: History and Literature in the Augustan Age* (Ithaca: Cornell University Press, 1994).

The invention of autonomy Historians of this period are often describing processes of eonic emergence without realizing it. J. B. Schneewind traces the complex chords of the discovery of autonomy from the rebirth (relative transform) of natural law theory and climaxing in the moral philosophy of Kant.[43]

Fig. 2.75 Beethoven composing

Perpetual Peace Kant is also the author of a famous essay on the emergence of an international system of peace, a text with traceable antecedents in the early modern, thence connected with the emergence of 'just war' philosophies. Alex Bellamy in *Just Wars* traces the tradition, appropriately (no accident!), to one eonic source, the Greek transition, "Between 700 and 450 BC, Greek city-states observed loose traditions aimed at limiting war…The Peloponesian War caused these customs to break down." A double eonic emergent! Note the concordance as to periodization of the Peloponesian and First World Wars. Note the pre and post divide timing. We must be wary of what we call an eonic emergent in this case, and be ready to refine analysis, since the appearance of 'jihad' in the wake of the Israelite corpus might also be called an eonic emergent, better in fact a degenerated mideonic echo. Our term, in this case, is too coarse-grained a sieve. Our model is too crude to solve the problem of war, indeed we see Hegel with dialectical precision fall in the trap with his remarks on warfare. At least we can be sure that our two-level analysis abstracts teleological unknowns from any connection to temporal drivers of warfare. Kant's thinking at the divide point sounds the clarion call for peace, most eerily in its timing. [44]

German Classical Philosophy Kant triggers one of the most remarkable surges of philosophical innovation in world history in the the *tour de force* sequence, Fichte, Schelling, and Hegel, concluding with Schopenhauer and Marx.[45]

---

43 J. B. Schneewind, *The Invention of Autonomy* (Cambridge: Cambridge University Press, 1998).
44 Alex Bellamy, *Just Wars* (Malden, MA: Polity, 2006).
45 Terry Pinkard, *German Philosophy 1760-1860* (Cambridge: Cambridge University Press, 2002).

A Riddle Resolved 213

Birth of Romanticism Our eonic model instantly transposes our viewpoint into a larger context where the issue of modernity is not the 'ism' of the Enlightenment, but the concert of many eonic emergents, among them the contrary descant of Romanticism. The sudden flowering of poets near the divide challenges the emerging scientism with a chorus of contrary poetic music.

The Pantheism debate Spinoza resurfaces from the early modern just at our divide and is reckoned against Kant in what is really the climax of the Protestant Reformation.[46]

Aesthetics With roots once again in the seventeenth century, we see the birth of aesthetics as a modern discourse, the contribution of Kant, once again, standing out in the birth of the Romantic reaction to the Enlightenment. Kant's third critique, paradoxically, almost has a greater influence than his first, in the reactions of Goethe and Schiller.[47]

Fig. 2.76 Thomas Paine

Bach to Beethoven…to Wagner In a mystery of aesthetic dynamics we see the clear relative transform we call 'classical music' peaking in the Enlightenment/divide period, reaching its climax at a white heat in the music of Mozart and Beethoven. This eonic emergent starts falling apart at the end of the nineteenth century.

Utilitarianism Our deliberate over-emphasis on a Kantian perspective should not for a moment blind us to the immense potential spectrum ('dialectic') of our divide period, seeing, for example, the parallel birth of utilitarianism as an unmistakable eonic emergent, perfectly timed. Our transition is a multidimensional set of innovations.

Adam Smith Seen in context, Adam Smith, perfectly timed, is a champion of liberty, prior to the emergence of the capitalism he senses coming into existence. Note how Smith has clear roots in the transition, e.g. with figures such as Mandeville.

The eonic evolution of evolutionism The idea of evolution is reborn in the

---

46 Frederick Beiser, *The Fate Of Reason* (Cambridge: Harvard University Press, 1987).
47 Luc Ferry, *Homo Aestheticus, The Invention of Taste In The Democratic Age* (Chicago: Chicago University Press, 1990).

Enlightenment as an obvious eonic emergent, and finds its first true theorist in

> **Lamarck** who produces the correct framework for a theoretical foundation of evolution in the double action of micro and macro factors...Darwinism will decline from this insight. This period also produces the teleomechanists, and the *Naturphilosophen*.

The period straddling 1800 periodizes as our transitional divide. The clustering of emergent processes is so massive as to be almost a dialectical flood. The transition to micro-action occurs within a half century.

> **The Great Divide** Our transition is swiftly accomplished and gives rise to the sense of a divide. Such a massively packed point of innovation is the best evidence of our eonic model.
>
> **Discrete freedom sequence** Like clockwork, 2400 years apart, from Solon to Tom Paine, the ratio of macro to micro-action spawns twice-born a democratic emergentism, just at a divide point. Now we see the logic of the mysterious timing of the great democratic revolution(s) of the end of the eighteenth century. Our calculus suggests that the divide line is the appropriate point for 'free action' to overtake 'system action' in the passage from eonic determination to free action, however 'free'. The brilliance of the generation of Thomas Jefferson passes quickly to the crystallizing outcome in the world of the Age of Jackson, as a new democratic experiment takes its chances as free micro-action in the new mideonic period. The Athenian experiment lasted about two centuries. The year 2000 might prove ominous for the American experiment.
>
> **Abolitionism** Out of the blue the abolitionists, appear just at the divide and the overcoming of the great curse of slavery is given its great historical first. The timing is almost uncanny, but our eonic model gives us the mysterious clue.[48]
>
> **Human Rights** A prime eonic emergent here is the concept of human rights which comes to the forefront in the eighteenth century, and along with it the (relative) transformation of concepts of natural law arrive just in time to stage an ideological accelerator for this period of revolutions.[49]
>
> **Feminism** A late-breaking eonic emergent (but we can see once again its sources in the seventeenth century), feminism is nonetheless another child of our transition, witness such figures as Wollenstonecraft, and its slow take-off in the nineteenth century will await fruition in the

---

[48] Eric Metaxas, *Amazing Grace* (San Francisco: HarperSanFrancisco, 2007).
[49] Lynn Hunt, *Inventing Human Rights* (New York: Norton, 2007).

A Riddle Resolved

twentieth.

Trend toward equalization We can stand back for a moment to see how misleading Darwinian thinking is. Evolution responds to the 'survival of the fittest' with injected trends toward equalization. Twice in our eonic sequence, beginning with Axial Age, we see the eonic determination

Fig. 2.77 *Peterloo* massacre

during phases of macro-action, of the evolutionary trend toward equalization. This emerges with unmistakable force in Rousseau, and we can see that the immediate tension arising in the contradictory new economic order. Equalization is an aspect of macroevolution.

Our transition draws to a conclusion with the great era of democratic revolutions, the passage to the new capitalism, the Industrial Revolution, as the nineteenth century begins the New Age proper of 'modernity', whose spectrum of opposites is a very balanced dialectic. Watered down renderings of secularism will tend to beggar this holistic totality.

The birth of liberalism From the seventeenth century to the point of the divide we see the gestation of liberalism, climaxing in its take off in the generation of the great revolutions.

The American Revolution It is hard to think of a more stunning eonic phenomenon than the almost uncanny and magnificent emergence of the great American democratic experiment, perfectly timed at the Great Divide, and showing the massive improbability of so many creative political 'revolutionaries', from Jefferson to Thomas Paine. A frontier effect inside a frontier effect, our transition seems almost deliberately to stage its novelty in the geographical fringe area of the open Americas, free of the inertias of European political continuity. The switch-off between system action and free action is clearly visible at once in the drop to a cruder lower grade, but essential, 'realization onset', seen in Age of Jackson. Simply spectacular.

Tom Paine Like Spinoza and Kant, Thomas Paine is one of the most perfectly timed gremlins of the eonic effect, appearing in perfect concert, as if with a task to perform, the clarion of secularism, economic freedom, and democracy. Dying out of fashion, in his wake the contrary tide of American fundamentalism will rise to claim a democratic revolution it did not initiate.

Age of Reason Paine's classic is accompanied by critiques of reason (reason noumenal or phenomenal?), and Hegel on Reason in History...

The rational the real? Our eonic model outflanks yet fulfills Hegel's classic rumination on the rational as the real, one destined to chaotification short of our rigorous division of levels. We see the eonic sequence expresses an ideal while mideonic micro-action may or may not be so legitimated as rational.

Industrial Revolution Revolution indeed! We tend to see modernity as characterized by capitalism, but this is misleading. Emergent capitalism is a classic 'eonic emergent' in the larger system of the modern transition. This 'relative transform' reinvents the already existing forms of commercial economy at a new level of technology and a new level of economic philosophy, or ideology.

The French Revolution to 1848 The same eonic characterization is deserved by the French Revolution, whose fate is to become the controversially ambiguous 'failure' of the period of the Great Terror. The democratic future will be endlessly delayed by the reactionary formations haunting the comparison with the American exemplar. The French Revolution also shows intimations of the nineteenth 'far left' emerging in the wake of the revolutions of 1848.

Tom Paine and the *sans-culottes* Paine has a close call with the *sans-culotttes*...The progression from the American to the French Revolution uncovers the latent contradictions in the liberal revolution as an eonic emergent as the element of class warfare enters with the birth of the step child 'socialism', and Graccus Babeuf's timely appearance at the first of the fake Thermidors.

Is there a Kantian Babouvism? The latent contradiction is expressed

A Riddle Resolved

Modernity, a New Age hijacked by Nietzsche

From LFM...
The Last Man of German Classical Philosophy:
Nietzsche, Darwin, and the *Genealogy of Morals*
http://en.wikipedia.org/wiki/On_the_Genealogy_of_Morality

Fig. 78 Nietzsche

Nietzche's position in modern thought is hard to untangle, and he is often subject to overvaluation: he comes at the end of the phase of German Classical philosophy and steals the show in a very dangerous way, the last man with a literary style takes all. This fellow is misleading, and his fans may never read Kant or, especially, Schopenhauer, the direct source of Nietzsche's theme of the 'will', but in a reduction of the 'will' to the 'will to power' and in a framework of scientism. Nietzsche's mistakes mixed with his genius make him an intoxicating 'poor devil' asking some good questions about the evolution of morality. But the answers he gives don't work, given the Darwinian context, and that of Lange, an obvious but strangely low-ball influence for such an intelligent thinker.

perfectly in the ambiguities of the classic liberal Kant's categorical imperative, and an antinomy of teleological judgment with respect to the 'end(s) of history', Babeuf to Marx, via Hegel.

Napoleon at Jena…Laplace whispers in his ear…Hegel…

The Restoration Is conservatism an eonic emergent? The incomprehending Burke, oblivious to his surroundings, nonetheless exposes the contradictory logic of revolution, as the drama of action and reaction play themselves out, from the streets to Paris to the Commune.

Romanticism…

Modern science…to scientism We have flipped the balance in our selection of eonic emergents away from the main event, the spectacular surge of modern science, toward the softer sounds of the multiple garlands of other emergent processes prone to being drowned out in the roaring thunder of the scientific revolution, cresting at the divide, onward through the nineteenth century. This temporary operational bias is easily corrected, and will itself correct our mesmerized focus on the science stream. This transition is almost overwhelmed by modern science, and yet, not. Kant with austere elegance poses the idea of freedom in a complement to the Newtonian triumph.

Fig. 79 Minerva and OwL If we are to look for initializations of religions at the 'divide' point, we have, in the mysterious humor of the macro effect, to look no further than Hegel and his mysterious *Phenomenology*. We should perhaps reinstate a cult of the *Owl of Minerva*, were not scolding historical materialists stamping our idolatrous reincarnations…

Schopenhauer The philosopher Schopenhauer, in parallel opposition with Hegel, produces a brilliant Kantian seed 'sutra' of superior quality to the decayed Upanishadism that will overwhelm Enlightenment discourse with another version of that term. The two neatly express a Buddhist and Christian line of realization.

Buddhism enters the modern transition in the wake of figures like Herder, Schopenhauer. Schopenhauer virtually reinvents Upanishadism/ Buddhism in his classic, beside the sublimation of monotheism in Hegel. This occurs just at the divide point!

Phenomenologies of spirit We have devised a means to outflank

A Riddle Resolved

The Great Divide

As noted, our eonic sequence is built around a series of short-acting intervals or transitions, and any such intermittent process will generate a 'divide', that is, the rough point at which the intermittent effect wanes and the outcome stabilizes. It is one of the most spectacular confirmations of our perspective that it uncovers this unsuspected aspect of the rise of the modern. We shouldn't be distracted by the secondary or exponential changes ignited by the new period generated. It is the core emergents, high-level cultural innovations, that are crucial, not their subsequent course. The downfield is something else. We deduce this in the abstract, and turn to our data to see if it reflects anything like this. It definitely does, and we can spot the right point immediately.

Thus, the period of the end of the eighteenth and the beginning of the nineteenth century foots the bill at once, and is one of the most fantastic (relative) 'start-up' periods of world history (a start-up inside a larger start-up, the transition), as the system crosses a 'divide'. This crossing point, a divide, comes near the end of the most recent of our eonic transitions.

In the space of a generation, the Dual Revolution of the English 'great transformation' of industrialism and the French political conflagration, as a volcano of the 'Left' passing into Socialism and Communism, initiate a global-scale 'crossing of the divide' that encompasses the American Revolution, immense cultural changes in politics, class structure, philosophy, religion, science, literature, indeed every category of human behavior. After more than two thousand years, democracy, driven by 'class struggle', emerges into universal acceptance after universal condemnation. The final assault on slavery rises with the paeans of Freedom culminating in the American Civil War.

Awash even after two centuries in a global transformation that dwarfs the memory of the wrathful minutes of revolutionary ardor in the streets of Paris, we arrive in our moment still animated by its momentum with enough distance to review its meaning from a greater perspective, and with an earnest hope, that only some phantom of the ultra-right could challenge, that as its children we will not undo its axioms. In a history of 5000 years we are barely more than a century past one of history's most terrible institutions, human slavery. And we would be deceived by our briefer time and the immediacy of a nearer moment if we complacently assumed that an action of Freedom guaranteed our future from the reaction of a greater time.

Hegelian metaphysics for an age of scientism, and yet we must pause to confess our wonder at the magnificent completion of the Protestant Reformation seen in its genuine 'prophet', the philosopher Hegel, and his version upgrade of archaic 'god talk'. This instant archaeological monument shows us an eonic observer first sensing the eonic effect, and giving expression, as did the creators of the Old Testament, to the eonic character of a transition in the eonic sequence.

Manchester…and the birth of 'socialism' The rushing logic of the modern transition shows the first signs of jackknife as the bourgeois revolution is sublated into a prophetically envisioned and renewed democratic revolution: a socialism of the proletariat, in a negation of the first outcome of revolution. The question of private property is too basic for easy revisions and the result will be the birth of a floating fourth turning point ideology.

1848: Marx, Schopenhauer,,… Was Marx a frustrated 'transcendental idealist'? The strange fissions of the 'Concept' show us two figures on opposite sides of the barricades of 1848, and it is strange that Marx's philosophy of history could so easily have been cast with a non-positivistic foundation. Wagner is there, and will attempt the perhaps failed, perhaps itself tragic, art-politics of the aesthetic state in his realization of his operatic labors.

Nietzsche Note how the philosopher comes very later after the close of the modern transition. Modernity ignited in the early modern could be called an immense flood of innovations hijacked by Nietzsche after its conclusion. He has created a great confusion by giving an immense artistic and inspired wrapper to what is little more than scientism.

We have garlanded just a few of the 'eonic emergents' and 'relative transforms' that characterize the modern transition. It is difficult to grasp the way so many creative individuals and innovations are clustered in the short rush of three centuries, with its climax at the point of the divide. We can see all at once that the explanation is eonic, and that such perfect timing reflects our frequency hypothesis.

System shutdown By the very nature of our model, we can see that the factor of macro 'system action', being intermittent, will wane and micro free action will rise to fill the void, with potentially ambiguous results. We see this effect clearly in the nineteenth century, despite its explosion of changes and innovations. The deep action of the early modern is at the source in almost every case. The dangers of chaotification or derailment are ever-present, and with the First World War and the

A Riddle Resolved

Holocaust we see the first of the mideonic calamities possible in this eonic progression. Take the measure of the modern transition: its action is at all points benign, then it stops. The continuations of completely uncomprehending politicians can wreak havoc in the outcome. Please note that scientism, Darwin, Nietzsche, come well after the divide point and yet rapidly purloin the definition of the Enlightenment.

Zooming in, zooming out We have done a kind of 'hundred yard dash' through the modern transition, culling a short list of eonic emergents, just on the verge of a more intensive look. We need to do the exercise many times from different viewpoints. We should, just here, before losing the forest in the trees, also zoom out to see the context against the backdrop of world history with just enough to see the clustering effect that once seemed like discontinuity but now seems like fullness.

2.9 A Riddle Resolved: The Macro Effect

Our snaphot of world history has uncovered almost without trying the presence of a non-random pattern of universal history by simple inspection. This pattern of self-organization can give us an empirical basis for considering the questions of human evolution. Instead of speculative theories like Darwinism we can discover a sense of universal history, thence evolution, purely empirically. To sure, 'facts' are seen from a particular perspective, but this doesn't alter the basic finding.

Our suspicion is confirmed that high-speed change can occur on the scale of just a few centuries, witness the Axial Age. And this effect shows us that evolution is hiding behind history in the form of a series of intervals of rapid emergence. World history yields its secret to simple periodization and shows from the invention of writing a clear developmental sequence, with a question mark about its probable source in the period of the Neolithic, the natural starting point for the rise of civilization. The great clue of the Axial Age suddenly provided the gestalt of a larger system at work. The Israelites were right, there is a process of greater evolutionary dynamism that frets the universal history of man.

> **The hidden structure in world history** We can call that sequence of three transitions and the epochs in between them the 'eonic effect', as a sequence of three epochs, and note the way that this pattern suggests 'evolution' at work, 'evolution of some kind'. It is at first illogical, it seems, to confound evolution and history. But with a little reflection we will see, first, that the two must be logically connected, and, second, that the

data we are discovering directly confirms that logic. This evolutionary sequence is a robust empirical foundation for understanding world history, in the context of evolution.

The relationship of evolution to history must resolve a paradox. The passage between the two could not take place instantaneously. It might show a series of transitional intervals that are evolutionary from one perspective and historical from another. But that is just what we are seeing: a series of 'axial intervals' or transitions that express a kind of evolutionary advance, and the epochs in between them that seem to express the historical carrying out or fulfillment of those transitions. What is remarkable is that we see this in historical times, and in a fashion documented by the rise of the technology of writing. It is futile to say that evolution must be purely genetic, since we can see that the 'evolution' of civilization is something more.

We have the first glimpse into the nature of human evolution: it is a larger process than the purely genetic development of the human organism. And we can see its last stages in the emergence of civilization. There are many more things to consider here as we proceed, but we have the basic insight into how we can revise our views of the meaning of evolution.

Note: A Certain Strangeness: Beyond Space And Time?

Our pattern of data has suddenly shown its resemblance to something remarkable and classic, so-called 'transcendental idealism', a scheme tailor-made to rescue Newtonian confusions, but considered now to be an outmoded form of thought. Almost against our will our model forces this on us, due to the two levels it generates in its analysis, and the stunning match to the discrete freedom sequence. Remarkably we have an 'off the shelf' philosophic software for just this situation, the critical system of Kant. In the next chapter we will tie together all the loose threads of our discussion with a look at Kant's essay on history. We can complete our model in the next chapter by showing how the eonic effect demonstrates the resolution of Kant's Challenge.

Our data has, at first, a strangeness to it in the way it treats discontinuity, jumps between periods and regions, and operates on fuzzy intervals. In fact, it is a consequence of the data we are confronted with, no way around it, and is not indulgence in the fantastic. Examine the data of the Axial

Age, for example. Fantastic or not, the data speaks for itself. There is no 'flat history' solution to the strange properties we discover there. One reason we are about to discover for this initial sense of oddity is that we may be detecting a system operating behind the scenes, and perhaps one that is beyond the matrix of space and time. Although we can't establish this formally, we should launch a preemptive strike against the suddenly metaphysical speculations that will arise here, and that will provoke some metaphysical spree on the subject of history and eternity. The latter concept has no scientific foundation, and is speculative, period. That doesn't mean it is wrong, only metaphysical. Transcendental idealism is the only way to both embrace and yet discipline this kind of 'ran off the meter' once we attempt to include anti-causal thinking in our model.

WHEE: 3.4.1

Note: The Kantian Age: Religion's Successor
_____

We are left to wonder at the future of religion in the modern age. The Kantian critique of metaphysics, poised between 'faith' and scientism allows a clarification of rapidly archaizing issues. The religions of the Axial Age are rapidly transforming in the modern world. But we see the rapid resurgence of vehicles such as Buddhism, with many New Age parallels, in modern secular culture. We should refer such issues to another essay, but contemporary buddhists have already created something very different from the religion of Gautama, or the Mahayana that arose near the era of Christianity.

The Reformation has created transitional vehicles that will either generate new religions of the future or simply dissolve into secularism. The new age movement has seeded a whole series of elements that will blend again into the secular. But the only candidate for the action we see at the dawn of Christianity is on the left with its revolutionary ideology. In fact the work of thinkers like Kant has transformed the basis or religion, and he takes Mt. Sinai to a new level of abstraction. After all of the drama of religion, the issue of ethics was never truly addressed. This happens for the first time with Kant and his trial attempt at a true ethical critique/construct with assumptions about will, freedom and non-theistic derivation via the faculty of Reason. Kant's thinking is theoretical and not a very useful moral canon in practice, as he notes by pointing to 'common ordinary morality', which is a complex of culture and innate logic. Consider the selection from WHEE:

Free Will, Moral Action, and Self-consciousness

The relation of representation and 'thing in itself' arises automatically in our model, and we should reluctantly admit as good materialists that this puts a big plus next to so-called transcendental idealism, wretchedly named. We merely noticed the arising aspects of historical appearance and the way this didn't quite add up, generating the characteristic turn toward transcendental idealism via the discovery of the uncaused historical intervals, 'freedom's causality'.

Our model posits outer psychological 'will' as 'self-consciousness' as the phenomenal aspect of 'will'...

We need something to tone up our discovery of 'freedom raw' in the enigma of what is clearly reflected in Kant's Third Antinomy. Like off the shelf software Kant's ethical thought, despite its immense complexity, foots the bill. Kant's ethical system is one of the greatest advances of modernity, yet suffers a faultline down its core, leading to a sort of gleeful Nietzschean reaction or spree pitting itself against morality

Our chronicle has temporarily skirted the issue of free will as a practical question by adopting a generalized framework of self-consciousness, in the contrast of system action and free action. All our account required was a 'self-conscious' agent with relative degrees of freedom or 'free action' in an evolving system action. His self-consciousness is the field of the manifestation of will. This 'free action' was not necessarily free will. This allowed us to construct a model that was deliberately fuzzy, and here neutral, as ersatz compatibilism (not the philosophic kind, but a simple fuzziness that is compatible with a deterministic or freedom interpretation

One aspect of the debates over free will lies in its timeless character. But we can see that our system might be evolving to the point where homo sapiens can begin to realize free will in action via his developing self-consciousness. In another sense, that potential was always latent in the potential of his evolved organism. In fact, we suspect, man always was, and is, 'ready' for this self-declaration. In any case, our model can easily do two things at once, and this corresponds to the distinction Kant makes between practical and theoretical reason. It is useful to stress this point since theories are not directly the basis for action. We should adopt the strategy that Kant urges on us, of making an operational assumption or postulate of the reality of free will, 'ought' implies 'can'.

WHEE 7.2.2

Kant also helps us understand the mysterious dynamic of the 'age of revelation'. The hopeless muddle of 'god in the gaps' resolves itself with simplicity and elegance and is implicit in the thinking of Kant on the antinomies of reason.

Freedom's Causality, Teleology and Politics

The inherent power of our eonic model exposes at once the basic resolution of Kantian perplexity. Kant predicts a teleological process he can't find, but which we have clearly found: 'freedom's causality'. As Elizabeth Ellis notes in Kant's Politics,

What would "bridging nature and freedom" mean outside of politics? For Kant the big questions are nearly always epistemological: thus, bridging freedom and nature might mean specifying the conditions under which investigators of the empirical world (scientists) are able to find evidence of spontaneity in the physical world (that is, of freedom's causality). Either freedom and nature are strictly alternative perspectives on the same set of empirical occurrences, or there are some things in the world that can only be explained according to freedom (in other words, the second alternative posits empirical evidence that some thing has no antecedent cause). I am not the first person to point out that it is not an easy thing to find empirical evidence of a lack of a cause. Kant himself assumes that a good scientist will operate under the presumption that absent natural causes may eventually be discovered.

But this is just what we have found, with respect to macrohistory, at least. The author complains that Kant's teleology and the necessity of free political action are in conflict. Our eonic model produces an independent teleological factor, visible only as directionality, that conditions but does not restrict human free action. Teleology enters our discourse as a perception looking backward of the eonic sequence, but this cannot directly change the nature of our freedom in the present, save to change the self-consciousness we bring to current action. That is, looking backward, we can see a teleological directionality, applying to macro-action. Our micro-action in its wake may or may not reflect that. This is critical for the preservation of freedom in history, for, as we examine the discrete freedom sequence, we see, remarkably, direct macro association with the emergence of democracy and this should lead us to examine the match to micro-action.

The necessity of assumptions of free rational action to conduct politics, conflicting with Kant's teleological thinking has been 'fixed' in our approach, by dropping the association of 'asocial sociability' with the driving action of evolution, and we can find the reconciliation of the contradiction, roughly speaking, in the way in which our two level system shifts gear between higher and lower degrees of freedom. This formulation allows us to free practical action from teleology, even as we allow this factor to remain in a larger system.   Adapted from 7.3.1 WHEE

2.9.1 Conclusions: Notes Ad Infinitum...

We have returned from a long voyage through the mysterious landscape of the Axial Age to find in modernity a successor to that period, and the dawn of a new epoch in the wake of a characteristic 'macro' transition seeding the onset of globalization. The solution to the enigma is to see the Axial Age in a progression of epochs, stretching back into the Neolithic or before, and constituting a phase of the 'Great Transition' as the renewed speciation of 'man' as *homo sapiens*. We hypothesize that Man's evolution is more than genetic and the working out of his larger potential proceeds through the cultural evolution of higher civilization. And the Axial Age is one of the most spectacular moments of that transition, one that gives us a glimpse of the 'evolution of religion'. Although the Neolithic is almost more 'Axial' as an 'axis' of man's history, the Axial Age as such is indeed a crucial turning point. It is almost an encyclopedia of potentials in the hypercomplexity of parallel emergent Greek, Israelite/Persian, Indic and Chinese and perhaps Mayan factors. But the real issue is the recursive driving activity of the sequence of transitions.

The point is to mediate the complex religious histories that spring from the Axial Age. We must note that Christianity/Judaism, Islam, and Mahayana arise in the wake of the Axial Age and that 'Israelitism' was an incomplete, tribalized, and finally discarded transition vehicle that was almost impossible to evaluate for the peoples of antiquity. But its classic literature took its place at the fountainhead of a new tradition. The outcome of two religions in collision was a teleological tragedy of a 'stream and sequence' system. But a similar 'jackknife' is visible in the outcome of buddhism, as Hinayana, and Mahayana.

There is an eerie analog resemblance between the Roman world and the American system (modern North America) as 'sidewinders' to transitions: the Greek, and the English. It is important to not be deluded by such comparisons: the element of free agency versus system action must include a reminder that 1. empire is a pathology of civilization, 2. empire, as the Romans well knew was a destruction of their great Axial innovation as republicanism, 3. repetition of the downshifting of the American system is already evident in its multiple 'imperialistic/capitalistic' episodes, but this is different from the question of the 'decline and fall' of the Roman Empire. The analog is closer to the collation of 'democracy' and 'imperialistic economy' of the Athenians who spoiled their great experiment within two centuries after Solon: by ca. 400 BCE the new politics was nearly spent. The uncanny timing here is like that of the American system, already dysfunctional after two centuries from its classic 'revolution'. The moral is not non-moral 'decline and fall' but that 'free agency could have/should have done better.

As we can see one of the great innovations of the Axial period was Greek democracy, which died out very early. A similar jumpstart occurs in the modern transition. But this is challenged very early with claims to 'real democracy' via social, post-capitalist systems.

Many statements about Axial Age innovations must consider the possibility that they were originally seeded in the Sumerian era. We don't see directly the first experiment in democracy, and much else, but they could well be there. The dismal cycle of 'macro' induction and mideonic loss of the innovation shows the reality of human free agency as it develops. The spectacle of human failure in democracy is dismal, and the further confusion with the process of capitalism is still another aspect of the loss of freedom.

In the model provided it is essential to distinguish progressive cyclicity which defaults to free agency after the initial 'punctuation' from cyclical recurrence which would be deterministic repetition. There is no dynamical logic to another round of 'empire decline' as we see in antiquity. The modern system is confronted with the difficult task of maintaining or improving on a system far beyond its technological/social capacity. And it is important to see that nothing in the gestation of the Axial period generated any causal inevitability to the process of decline. The whole point of the macrosequence is to jumpstart human cultural activity at a higher level obviating repetitive decline.

We should expound the 'secular' as the receivership of ancient religion, but this must be done without the reductionist scientism that has become temporarily dominant in 'modernity'. We must finally adopt the phenomenological foundations of our model and suspect the noumenal limits to our perceptions. The moral of the elder tale is becoming that of the newer modern, as we are being driven into the future, into a new era, and our ability to correctly observe and assess the Axial Age is already limited by the passage of time and the entry into a new world. The data of the Axial period has been so abused by monotheistic religious myth that is very hard now to deal with what should be a unifying historical theme appropriate to an age of globalization. All themes to the effect that 'god acted in history' via the transition of Israelite Canaan are obsolete now, and would have to be consistent in any case as to similar claims as to the vast structure of synchronous parallels. We could hardly claim that buddhism is the action of Jehovah in this 'age of revelation'.

> Our data produces a remarkable stalemate as to 'design' arguments, and this is pursued at length in the Conclusion. The perspective of secular humanism and Biblical Criticism fails because a clear 'design inference' is inevitable, but the Biblical 'design' argument fails because the 'Jehovah' myth as pop theism also fails...the notes attempt to discover the concepts needed in this 'crisis' of theology,...and of theory.

> We can be atheists or theists but we can't so easily dismiss the possibility of spiritual powers in nature. We must distinguish life, consciousness, will, in the manner of Schopenhauer, to have any hope of making sense of 'something inside existence' that is beyond life yet conscious and willful in some sense. With this approach we see at once many of the confusions of archaic theology for what they are.

The rise of modern atheism is a powerful counter to the Occidental monotheism generated in the Axial Age, and began early in the modern transition, but the secularization of the cult of 'Jehovah' is not the same as the one-way valve of the Axial impress of 'meta-theism, which is likely to survive its first religions, reflected in the mysterious fragment of a lost legacy of the 'unspoken name of god': IHVH. The point is clear in the way the tide against paganism and 'idolatry' remade the ancient Roman oikoumene, followed by the complex and still hard to understand Islam. This is seen also in the way that modern atheism attacks the de fact idolatry of 'pop theism' and its rapid ascent as a new 'superstition'. The dialectical interaction of theism/atheism comes to the fore in figures such as Kant and Hegel, and the philosophy of Schopenhauer as an inverted theism is a brilliant counterpoint to both the Reformation and the rise of Secular Humanism. 'God' as 'being' and 'god' as 'will' would really require two different forms of atheistic negation.

> One of the 'innovations' of the modern period is the 'transcendental idealism' of Kant and Schopenhauer, which greatly clarifies the confusions of antique religion/monotheism. But this was clearly anticipated by Plato! The Axial Age again! The absolute idealism of Hegel becomes a riddle for the future, unless his 'phenomenology' explains the duality... Note the appearance of this at the modern divide! Right on schedule...

> The modern system in the early nineteenth century jumps the track into

a fundamentalist 'materialism, positivism, scientism'. This has actually arisen in the wake of Newton and was latent in Spinoza and much of the fanaticism of the 'new physics' and is a 'free action'/ 'system action' effect as the quality of the system declines rapidly in the wake of the transition...Kant pointed to the corrective to see a fuller modernity in the 'dialog/dialectic' of causality/freedom. We must suspect the same in the ancient transitions: what we see may already be a diluted outcome... The correct interpretation of 'modernity' is very complex. But this new track may be the prelude to scientific versions of materialist 'Samkhya' in the elucidation of 'consciousness' and 'evolution' closer to our usage and empiricism.

Our historical perceptions have expanded dramatically, but we lack the right conceptual framework to analyze the phenomenon of the Axial Age.

We have confronted a problem that the primordial *Samkhya* ('the spiritual + 'god' has weight and measure', as higher Maya??) dealt with: the material or phenomenological factor of 'higher spirituality'. We hardly know what that is, but the Axial Age gives us many examples!

The rise of modern historiography and the discoveries of archaeology have put the image to fixer and we begin to see the outlines of a stupendous phenomenon of almost *Gaian* proportions in a lightening flash across Eurasia from Rome to China (with the Mayan case an hypothesis, and the case of Africa conforming to the dynamics of the 'frontier effect'). Note that 'Europe' is a fictional subset of Eurasia, not a separate civilization, and shows the diffusion from Neolithic sources which we suspect form the earlier stages of our larger sequence addressed to explain the Axial Age as a stage in the evolution of civilization.

> Europe shows thus no participation in the Axial phase, and will be a prime candidate for a frontier effect in the spectacular rise of modernity in an untried zone. The case of Africa requires seeing that it has long since joined the macro assembly and is still in the process of the expansion of Neolithic cultures throughout the sub-Saharan continental mass. We should keep in mind that the interior of African was almost impossible to even explore until the rise of modern medicine and its penetration in the millennia after the Neolithic must have been a slow adaptation by these (e.g. Bantu) missionaries of higher civilization.

The association of the Axial Age with monotheism is misleading, as we see, and the emergence of at least two (with Zoroastrianism and Confucianism at the side), parallel world religions in the wake of 'Israel/ Judah' and the Indian Axial Age, theistic on the one hand, and atheistic on the other, must remind us that our categories of analysis are barely adequate to the study. And this becomes the more complexified in the perception of the Greek Axial Age, which in many ways spawns the modern secular world. We do not really see the sources of 'monotheism' in archaic 'Israel/Judah' and the subsequent obsession with supernaturalism of the religions emerging in that sphere should not let us forget the stealth 'religion of freedom' seeded in ancient Greece (if not ancient Sumer) and this legacy, sinking like a stone in the medieval era, resurfaces with a vengeance in the modern transition. The association of the 'secular' with scientific reductionism is thus misleading and indeed at the divide period of the modern transition we see the shift to a new philosophic successor to religion, and the careful critique of metaphysics by the efficient 'non-prophet' Kant, midwives this new 'post-religious' consciousness of the dialectic counterpoint to the New Physics of the idea of 'freedom'.

We were thus reminded to include the spectacle of Archaic Greece in our analysis, and its prophetic incidents include virtually all the seeds of the modern outcome, via its gestating Scientific Revolution, its birth of philosophy, its epic/tragical poetic genres, its incipient historiographers, and its classic first springing from a field of proto-republican experiments of a first democracy in a spectacular yet brief manifestation in Athens of the 'freedom' nexus, *eleutheria*. The larger picture includes an immense wealth of realizations, from Confucianism and Taoism, to the Indic Vedanta and Upanishads, its Jain Teertankers, and its global 'world religion' of Buddhism. We tend to miss the emergence of two monotheisms, braiding at the period of the Exile, that of Persian Zoroastrianism next to the cult of YHVH, later the theism of 'Jehovah'.

> Much of this requires a careful understanding of our distinction of 'stream and sequence' and we see at many points the way a stream element is taken up and amplified in the 'sequence' transformation of the Axial interval, the prime case being monotheism itself. It is difficult to sort out the dynamics here without this element of our model, for example, the 'double' birth of monotheism, first as a primordial version (as seen in the Abrahamic legends), and then as a transformed, and amplified Axial Age production. Over and over we see this trick played on the elements of greater antiquity in their Axial Age recursions.

Our model proposes a distinction of stream and sequence which clarifies at once a host of historical confusions and mysteries, e.g. the way monotheism seems invented before being invented in the Axial period:

Stream and Sequence Over and over again our distinction of the stream and the larger sequence clarifies the confusing layers behind the Axial

A Riddle Resolved

Age (and the earlier and modern cases):

Israel/Judah the saga of 'Israel' as a longer history AND a short interval of an 'age of revelation' shows that the 'Israelites' were beginning to understand the distinction of levels. Everything about the Old Testament revolves around two types of material: the larger mostly mythological account and the short period from the period after Solomon to the period around the time of the Exile.

Persia The stream aspect of Zoroaster bequeaths a version of monotheism, one that is blended with the empire phase of the Persian world in the classic core: the frontier effect with Israel carries the day as the two blend at the end of the 'transition' interval.

India that historical evolution is more than factor causality in a temporal succession can be seen from the case of India where three separate entities become the objects of effect: the Upanishads and their legacy of 'Vedanta', the Jain tour de force of 24 Teertankers climaxing in Mahavir, and the emergence of the remarkable 'buddhism' as a world religion in parallel to the Israelite action.

China The stream aspect of the Chinese historical legacy is very strong and it is hard to even detect the relative transformations of a 'macro' interval, and yet right on schedule we see the almost uncanny birth of Confucianism, and Taoism ca. 600 BCE!

Greece The Axial era in 'archaic' Greece (which spills over the divide into the spectacular 'classic' flowerings, two centuries to ca. 400 BCE) is the *piece de resistance* of the whole Axial Age, and gives new meaning to the idea of 'revelation' with its massive innovations in multiple fields…

(**Maya?**) The isomorphic synchrony of the Mayan Axial period remains a mystery but should be included on provision, given the lack of exact data. The 'stream' aspect of the Olmec, as per our model, is of course quite different. Cf. our discussions in the *Preface*.

Our suggestion of a 'macro-sequence' in a frequency of 2400 years exposes the last three terms of a possible progression, with a 'mini' Axial Age in Egypt, Sumer:

--7800 BCE ?

--5400 BCE ?

--3300-3000 BCE Egypt, Sumer…?

--900-600 BCE, Eurasian spectrum, (new world?): Greece/Rome (Rome is really like one of the 'city-states' in the Greek constellation), Israel/Judah, Persia, India, China

--1500-1800 Eurozone frontier effect: the crescent, Germany, Holland, England, France, Spain, (north Italy??), around the frontier of the old Roman Empire, a stunning example of our model...

This model exposes the highly suggestive period hypothesis ca. 5500 BCE as strong sequence source for an emerging mainline, with a question mark about an elusive 'proto-Axial Age' seeding some the of source proto-religions that reappear much later: a good example is the mystery of the source of the remarkable Indic religious legacy of 'primordial Shaivism', the source of the later yoga/tantra/samkhya and the ancestor of buddhism in Jainism. Hinduism is a later hybrid of this earlier tradition and Indo-European elements.

This obscure pattern makes sudden sense using a model of a cyclical driver in a frequency 2400 years with a 'minimax' process that is teleological in a special sense: unidirection must balance lateral diversity. During the Axial period 'direction' idles as the lateral is the object of a spectacular synchrony (although the world religions attempt to usurp direction), and then 'direction' is reset outside the given system in the extreme jump to the Euro-zone at the far end of Eurasia! Clever indeed!

This model explicates and replaces the hopeless muddle of the ancients over cyclical views of history which have been also mixed with astrological superstition which has survived even in J. G. Bennett's *The Dramatic Universe*. The myths of 'new ages' are false but a similar 'new aging' follows a different tempo while our systematics and causality resolve the confusions have nothing to do with cosmology. But it is not clear if this macrosequence is active in the Paleolithic.

Much 'new age' thought, confused further by 'postmodernism', attempts to perpetuate the ancient lore of the 'new year' cosmology (e.g. *The Aquarian Age*) and is hopelessly out of date. See WHEE 6.6 *New Ages*, 6.6.1 *The (Eonic) Evolution Of Religion*, 6.6.2 *The 'Axial' New Age*, 6.6.3 *The Great Freedom Sutra*

We must also consider our analysis of the 'frontier effect' to see how 'Israel/Judah' and 'Archaic Greece' show the way our 'macro' effect always restarts in new zones, but adjacent to the prior phase. Why, after the immense phases of civilization in Sumer and Egypt would a new era bypass these developed areas to start in a backwater like 'Israel', or in the wilds of archaic Greece? Our model codifies the rationale.

Frontier effects Our model is based on a 'frontier effect' whereby the mainline jumps to new adjacent zones in its next phase: the mainline operating via Sumer is halted by the sudden splitting across Eurasia, with new zones: Greece (Italy), Canaan (Israel/Judah), Persia, India, China (but nothing in Europe!) and ?, Mayan civilization. All of these 'upstarts' in parallel are really in the prior Sumerian diffusion zone, which reaches as far as China, Europe, and the New World. This is the precondition for a next step. Note the elegant way 'Israel' and Greece are frontiers in the new era. Note the way the frontier zone of Persia is compromised by the way it seizes the core zone (once Sumer) and gets its contribution diffused at once into the Israelite! The modern case will be different: the global system must resume a mainline and this is in a frontier zone in extreme Western Eurasia! A barely comprehensible but very elegant novel system for directional/integrating globalization. Since we only have three beats in sequence, we must expect skepticism, but the flat history approach is so incoherent, and this simple model yields so much insight, that we suspect we are on the right track. What we have is

>3300-3000 parallel 'axial' Sumer/Egypt...?
>900-600 parallel 'Axial Age' list
>1500-1800 a subset of European sector...

And this frontier effect explains at once why the Zoroastrian equivalent of IHVH monotheism never quite gets off the ground: its source in Persia is a frontier effect, yes, but this as it becomes dominant moving in the core zone of Sumer, sputters, becomes an ideology of emperors, and ends by blending with the Israelite brand during the Exile. Our larger system is rooting for underdogs! Remarkable, and the frontier effect at its most confusing. But this legacy will and probably explains why a second monotheism will spring up much later in the new era: the Islamic parallel to Christianity, steeped in the complex mutations of Zoroastrianism. And the whole Axial spread across Eurasia is another *de facto* frontier effect (although technically something different in our model) in the parallel independent emergence of transformations in India and China. Africa shows no Axial effect: it was one of the sources in the era of classic Egypt.

The religions of India are a mysterious complexity and the analysis is complicated by the confusion over the much debated 'Aryan invasion' of that subcontinent. Many regressive claims are now proposed to claim that there was no such invasion and that the Aryans actually sourced in India itself. Such claims invariably miscalculate the evidence of linguistic similarity and evolution. The parallel of Vedic Sanskrit and Homeric Greek is too close for any attempt to refute the 'Aryan invasion theory'. In fact we suspect the great antiquity of Indian religion that preceded the Aryan interaction, if not invasion.

We suspect that much of what we see later springs from the Neolithic period, and the cults of 'primordial Shaivism', the ultimate sources of yoga/tantra, *Samkhya*, and Jainism. We are thus confused by the historical blending of Indo-European traditions with indigenous ones in the misleading tradition of Vedism. It is thus remarkable that the Axial interval instantly uncovers the real tradition and we find the Upanishads at the dawn of the axis point, followed by a Jain phase, followed then again by the birth of new global vehicle distilled from the greater tradition, the early buddhism which begins its epochal career in remarkable concert with the Israelite. This spooky timing of theistic and atheistic religions is a challenge to both religious and 'secular' interpretations. The ability of our macro processsor to sort out and separately process three independent strains, the 'hindu', the 'Jain', and the 'Buddhist' to come, is strong evidence that standard sociology isn't going to work here!

In the notes to the Conclusion in chapter three, along with the remarks in the *Preface*, we have tried to sort out design arguments, mechanical models, issues of teleology, with an idea from the ancient *Samkhya* no less of the higher 'spiritual materialism' undergoing transformation such as we see in the macro aspects of our data and model. It is a highly elusive and still unexplained mystery and one that has rendered the theologies of 'pop theism' obsolete. Any examination of the modern transition is an overwhelming reminder that this onslaught of critique is the case, and we reiterate the obvious that the religions of the Axial period are under assault by the rationalist analysis of modern historical sociology, itself rendered too often dysfunctional by its reductionist scientism, and its confusions over the both of evolutionary theory we find in darwinism. Modern science has at yet no categories for the correct analysis of the Axial Age, and the methods of conventional historiography have completely failed.

After all, the idea of an age of revelation was not so far off, however primitive the confused and remythologized account of the Israelites. But we see that the Greeks' age of revelation from the archaic period was almost more remarkable and further the failure of the Greeks to observe what was happening ironically made their testimony the more convincing and untampered with. The pristine evidence of an interval of axial transformation is thus almost the better preserved in the record from the time of the Homeric period to the last Euripidian moment at the end of the Peloponesian War.

But to be fair, the Israelites so-called performed a remarkable first: the observation as it was in motion of an axial interval in transformation, and it not surprising they thought this could only be the action of a divinity. The cult of this divinity rendered into a tribal god became the output of the transformation in question. In fact this is a record of macroevolution of religion in action and the overall account is a true first at the dawn of written records. In a further irony the modern so-called 'secularist' will inherit this account and savor its real meaning more directly than the religionist and the primitive record will be become a treasure of the historical record of macro transformation in the evolution of man.

The real conclusion to our 'story' is the successor epoch to the Axial Age that we call modernity and its reformations, scientific revolutions, its political revolutions, its art and literature, and its ambiguous economic realization. The latter is already beset with precisely the kind of 'revolutionary' chase plane formation we see with the pursuit of the Roman empire by the religion of Christianity: the challenge of postcapitalism.

The future of the world system thus remains in question and we can see that much of the legacy of the early modern is already fading. The issue of globalization and the coming of postcapitalism is the unsolved riddle posed in the wake of the French Revolution, and those of the year 1848. We should note the definite resemblance of the two 'sidewinders' to the (Greek) Axial and modern transitions, the Roman and the American (United States), suggesting the recurrence of the 'empire' mode to produce a larger oikoumene. We should refrain from the temptation of thinking history will repeat itself, and not sentimentalize the grossly regressive catastrophe of the Roman barbarism, and we can see that the seeds of an alternate avenue came into existence with the challengers to capitalism. We should ask if democracy will survive, or pass away as it has before. The issue of climate change has accelerated the time table of 'mideonic' failsafes, and the formal analog of the succession to the Axial period is going to be misleading.

We are forced to leave behind the Axial Age, and we don't have the conceptual tools to really solve the riddle. Our partial solution leaves the larger question of evolution, that of man, and the rest, only partially answered. Part of the answer lies in the analog of the modern transition, but ironically that is also become a mystery in the age of 'scientism' and the strange distraction of darwinism. As man moves into the middle era beyond the onset of a new epoch he is confronting with the scale of his 'micro-revolutions' confronting the 'macro-revolutions' of the large 'eonic sequence'.

The nature of long range change in civilizations is the challenge of man's hi-tech, and hopefully, hi-culture tech, future, one beset with the mystery of spiritual powers in nature, beside the question of alien worlds. The former, one suspected aspect of the transformations we see in history, is something far beyond the issue of encountering alien life, if any, and impinges on the 'spiritual realm' inside nature, a world man is entering in enigma of his own consciousness. As man passes beyond the religions of the Axial Age the danger instead is of a reversion to barbarism in a slide against modernity itself. We must note the eerie timing of Athenian democracy, gestating with Solon, yet gone by the early fourth century. Beside that we see already the transition of the now classic, and only, that moment of the American attempt, already taken over by conspiracies of the spy film specimens.

We are in the first period of world history when the mystery of historical evolution could find a solution. Our basic 'pattern' of five thousand years arose first in the nineteenth century and the phenomenon of the Axial Age was noticed at once. And then its larger context. We have provided a hopefully falsifiable hypothesis as to a still larger pattern and context and we will see if our generalization can be empirically confirmed. But the loss of close range data makes the task difficult. That is the remarkable irony of the great achievement of the 'Israelites': to see the macro effect in action and leave a record, however primitive, of a macro transition. The confusion of the 'output' a monotheistic religion to challenge polytheism, with the dynamics of that very transformation leaves us now with a remarkable insight into the now lost epochs of human evolution when the challenge of time and memory produced myths of intervening divinities along with legacies committed to oral transitions. In that sense the Israelite transition was a modern innovation with its new hot technology of writing! The real issue of the 'Israelite' transformation is thus not religion at all, but evolution.

It is hardly an accident that the modern period was rapidly flooded with the diffusions of ancient Buddhism, and it is the reality of the 'frontier effect' that the Euro-focus of modernity forces the issue of 'life-boating' important legacies of antiquity. Buddhism with its carriage of the lore of human consciousness and its transformation is one of those, and the mysterious world of Sufistic Islam is another. But the antagonism against modernity of

A Riddle Resolved 237

many of these now vagrant traditions will force them to seed and 'fall into the ground and die', as the new era almost automatically recycles ancient leftovers of the Axial period.

One of the ironies of modernity is that the early modern transition as a successor Axial phase is so complex we can barely understand it, in its variety and depth, and the result is that we have become fixated on technology at the expense of a larger vision at work. And it is important to see that capitalism appeared very late, at the end of the modern transition, and in concert with an antithetical challenger, communism. The threat that capitalism will simply swamp the whole future of man requires the economic perspective be seen in its real context. And we can see that democratic government is rapidly being taken over by Machiavellian politics, the psychopaths of the spy film, and the imperialistic economics of the period of globalization. It is important to remember that the advances of modern politics were degraded, not advanced, by the Machiavellian mindset. And the mysterious way that Nietzsche neared hijacked modernity.

Our fate is to be preoccupied with modernity, a sort of 'second' or 'nth' Axial Age, and we are already watching the world of proximate antiquity slip into empirical ambiguity. Ironically, the one phase of the Axial period we can really say is empirically 'image from fixer' is the Archaic to Classical Greek period, and is is obvious that modernity shows the resurgence of this contribution, with, to be sure a Reformation to overlay the Judaic on the secular, and, to a close look, entries of the Chinese tradition, and the sudden flood of 'new age' diffusion starting at the time of Schopenhauer who virtually reinvented yogic psychology for the modern age.

> After all of the drama of religion, the issue of ethics was never truly addressed. This happens for the first time with Kant and his trial attempt at a true ethical critique/construct with assumptions about will, freedom and non-theistic derivation via the faculty of Reason.

The modern period in the wake of the 'divide' shows a sluggish start as the immense complexity of the early modern resolves to the era of the Iron Cage. And the slot given to religion in the wake of the Axial period defaults to the world of the leftist revolutionaries and their novel materialist bent. A new age indeed. We should note however that the first initiative of modern communism was that of Thomas Münzer at the dawn of modernity. But we also see the great recursion via the modern democratic revolutions of the first experiments in democracy from Archaic Greece.

There is the prospect of coming to an understanding of the Axial period via the study of the modern transition. But the context of modernity has quite obviously shifted as the term 'secular' both explicates and confuses the issues in the attempts to define it via 'scientism'. The modern period shows the equal shift to economic categories in the spectacular emergence of capitalism/communism just at the divide period, and we see the same trick of emergent 'chase plane' movements, in antiquity monotheism, in the modern phase post-capitalist challengers to the economic frenzy of globalization so rapidly become problematical as a threat to the entire planet.

The modern transition strongly suggests that a long phase of 'macroevolution' starting in the Neolithic (we suppose) is coming to an end and that man is now left, or bereft, as he looks backward at a history of evolution in our new sense, now concluding with the increased degree of freedom of the men at the outcome destined to need some understanding of the Axial Age so he might replicate such actions in the revolutions of the future. It is a freedom fraught with peril and we should doubt if it will arrive via 'free markets'. It is the tale of the first men to emerge from the last men, in a possible reference to Nietzsche, or else to first and last hominids, the first of species of *homo erectus*. Man as incipient homo sapiens has not completed this transition and the case of *homo erectus* is closer to the reality of mechanical man than we might think. As Wallace realized, man has a potential as yet unrealized, to come via the freedom to evolve, or self-evolution of man in the freedom of his own becoming.

Art, Evolution and The Tragic Genre

We are confronted by the fact that Greek tragedy arises in the Greek Axial interval, flowers in spectacular fashion and in perfect correlation, then begins to wane promptly at the conclusion of the transitional interval. In terms of our evolution formalism the correspondence is eerily exact, in terms of macro and micro, System Action and Free Action. We are left to wonder about earlier stages of human evolution if we see such spectacular kibitzing at the level of art.

Thus, the historian William MacNeill, in *Keeping Together in Time*, considers the element of dance and song in human evolution. But this process is right under our noses if we carefully do some accounting of relative transforms in our eonic pattern. Most 'song and dance' elements are well established in the human legacy and cease to show relative transformation. We need to find one that is inside the eonic mainline, what we will call an eonic emergent. We can see that the eonic pattern is pervaded by spectacular cases of artistic flowering. Here is a prime case for our distinctions made between what is potential at all times and what appears in our macroevolutionary pattern. We can in fact isolate one spectacular intermittent effect in the genre of Greek tragedy (whose 'song and dance' elements are almost vestigial, as it passes into a literary genre). Its relevance to our 'evolution of freedom' is direct. And the suspicious similarity of the 'tragic theme' to the issues of religious evolution should alert us to the importance of the issue. The potential to create art, acts of purpose, and will, and the freedom to 'screw up', closely resemble each other. This is a complex subject, but our remarks will be restricted to periodization, and it also true that the example of the tragic genre, although of special interest, is only one of a whole range.

As we move to create a model we need to remind ourselves that aesthetic issues are a still more complex domain beyond even the ethical ones we find lacking in causal thinking. Later we will look at the philosopher Kant, and there find it no accident his Newtonian musings split into three critiques, one each for the causal, ethical, and aesthetic modes, with an ambiguous fourth as to the teleological. As a token of the complexity of (eonic) evolution we can notice the issue of the evolution of art embedded in our data. Note that, from a high-level view, seen in retrospect, we can see that as the Axial interval switches on somewhere ca. -900 a whole series of literatures start coming into existence, accomplished by -400 at the latest. Nothing in this preempts later contributions, but the relative effect is unmistakable, occurs simultaneously in five or more areas independently, and shows feats never matched even today. Note especially the sequence from the Iliad to Greek tragedy, which suddenly appears very briefly. This kind of data is beyond analysis in current science, yet simple periodization forces a paradox. We are approaching a crisis of analytical concepts. The difficulty of the tragic genre makes its appearance ultra-rare, and as it happens it sandbanks inside our pattern.

In general, let us note that our 'evolution of some kind' seems able to leave great art in its wake, as a matter of relative transformation, i.e. in the intermittent series visible as the eonic effect. Please note what we mean, the potential for art already exists in man and occurs in every generation but at a relatively higher degree of contingency, the random distribution of genius. Here we see our 'evolution' inducing a spectacular clustering period of the highest art, e.g. Greek Tragedy, with or without the factor of genius, against (to some degree) the element of contingency. Later periods can't continue this because they don't understand it.

This 'evolution' doesn't just generate art, it generates relative transforms seen in periods of higher, the highest, level of art. Yet human creativity is never violated. We know this only by periodization and careful accounting of time periods. Therefore this 'evolution' operates at some higher level than the highest level of art. The same could be said of philosophy or religion. Shall we go on? Darwinian stock is starting to collapse. We have several million years of coarse-grained observation of Darwinian evolution, and five thousand years of fine-grained observation of some other 'evolution'. Are the two the same, or did one pass into the other, and if so, when? WHEE

3. CONCLUSION: THE GREAT TRANSITION

> ...Since the philosopher cannot presuppose any [conscious] individual purpose among men in their great drama, there is no other expedient for him except to try to see if he can discover a natural purpose in this idiotic course of things human. In keeping with this purpose, it might be possible to have a history with a definite natural plan for creatures who have no plan of their own.
>
> *Kant, On History...*

Our exploration of the Axial Age is complete and we find ourselves confronted by a remarkable mystery but with a partial resolution using a new form of historical model. And this tells us to look for a sequence of Axial Ages, and that one of the best clues is to examine modernity itself, as though it were a secondary or later 'Axial' period on its own. This model might seem *ad hoc* or artificial, but its method scores a bull's eye and is entirely appropriate as long as we don't confuse 'systems analysis' with a 'causal analysis' pretending to be science. And this approach can shift gears between such causal and other forms of explanation, including even purported 'spiritual' design accounts. However, we will in this conclusion attempt to move beyond the model with some ideas that are speculative, but which can ask some new questions about the enigma we have found. This is against the grain of our Kantian perspective on metaphysics, but once that is understood we can incrementally increase understanding with a survey of some possible solutions to our enigma in a novel form of that very genre, metaphysics. In fact, the twin solutions, biospheric nature, and a demiurgic manifold, are potentially 'best of luck' empirical issues open to some form of demonstration, to a future science. They are not issues of 'faith'. We will explore these issues in a very repetitious fashion, since the concepts are unfamiliar.

> The biospheric interpretation fits, is 'neat', but invokes a speculative *Gaian* 'myth' we can't fully resolve, save to note that pace Bennett the 'biosphere' has 'will' and hyper-consciousness, and transmits some kind of cosmic imperative of 'life on planets'. That's not adequate as explanation we admit at once, but bypasses theistic historicism which breaks down with the complexity of Axial Age data. Both of our concepts, demiurgic and *Gaian* biosphere should probably resolve to an undiscovered aspect of nature, e.g. some brand of (cosmic/quantum) computation and spooky physics.

Our data is so spectacular that our model might seem to fail to do justice to the spectacle of spiritual history, so-called. It is hardly surprising that the Israelites thought in terms of theistic historicism. They pointed correctly to a 'phenomenon', their own history, that defied common sense history. They were right, but we should grace these 'primitives' with the badge of scientist with a falsification of their clever interpretation which can't see the clear signs of mechanization of that history. For example, why does everything revolve around the obvious evidence for our model of a 'divide' around 600 BCE? Why would 'god' operate over a finite interval or in cycles? Or produce atheism and theism in parallel, in our assumption via the model that synchronous phases show a common determination?

Our larger dataset is in many ways far more spectacular still and a cardboard divinity like 'Jehovah' seems primitive now. We can't rule out a hidden entity plying a fake divinity for a tribe of primitives, but in the end the larger spectrum of the Axial synchronous transitions finds better, and in many more spectacular, account of our 'systems model'.

> We should offer a disclaimer: we have found everything ancient minds thought was 'god in history', but we can see that it isn't 'god'. However, we cannot prove either atheism or theism, if only because we can't even define 'god'. Our 'edge of space' antinomy shows what the ancients must

Conclusion 243

have thought was god: the antinomial mystery that was like a larger dimension, and this was present to all points in space, as would be a hyperspace, god closer than your jugular vein, as Moslems put it. If we call this 'god' we should be wary of thinking this would intervene in one isolated corner of the universe, and one isolated corner of the planet, in one isolated interval. Our model makes much better sense, but what runs this model?! Modern philosophy has outlined the three possibilities, materialism, absolute and transcendental idealism. The latter suggests we are constructing our own space-time, which is a joker in the deck. The 'materialism' of *Samkhya* makes these distinctions seem artificial. All three could be true. The sense that this superset was the source of the 'will of god' became the ancient version of pop Schopenhauer. We should debrief this, but we can't debunk it because its antitheses are equally antinomial. Here 'faith' is equivocal, and the Buddhist Void works just as well, if not at all. We must leave it at all, with a solid critique of the Old Testament brand. But the latter will fall into the lap of the modern secularists as a richer mystery than the original idolatrous monotheism. An archaeological site in the classic history of man. It is a secular inheritance now, with the prize going to the Israelites for first detecting the macro effect.

Nothing could be simpler in the sense of our model: the data of world history is highly suspicious and shows a clear non-random pattern, so we test the data with a frequency hypothesis at various intervals and find surprisingly strong evidence of a cyclical system with a driver generating a mainline, with side effects operating in parallel. We have used an 'evolutionary' formalism for this. But the sense of design is very strong. To criticize our use of the term 'evolution' is beside the point, it means 'brown paper bag' holding data on 'development'. If the data shows in addition two levels allowing a macro/micro distinction, our usage is still more appropriate, if not to Darwinians.

It is hardly surprising that the Israelites thought they saw 'god' in history. The data satisfies a 'design inference', a very strong one. Try to explain the twin mirror image of two monotheisms Zoroastrian and Israelite, Aryan and Semitic, blended just the point of the Exile, our 'divide'. No hypothesis of linear causality is going to explain this, among many other, 'spooky' complex incidents. But we must move with what our model shows us: the clear mechanical character of the data, but 'mechanical' in a new and mysterious sense we will explore in the notes. We need to adjourn, with a lot of notes, in short order, with a study project for a spectacular phenomenon rather than a metaphysical 'spiel'.

The Axial Age can escape from us: we see six parallel immensities we have barely explored from the birth of the secular in Greece, twin mirror monotheisms, Persian Indo-European and 'Israelite' Semitic so deftly blended at the period of the Exile, a manifold of Indic religious manifestations climaxing in a Buddhist externalization in the wake of the Axial Age, and the twin manifestations of Confucianism and Taoism. And we must note the larger sociological evidence of the Greek instance: we see all sorts of fancy innovations, but we also see a cultural evolution, simply because we have more data. We don't see that properly, for example, in India, and focus on Buddhism, for example, with no sense of the cultural background. But perhaps our 'smart system' is so smart it can override these issues and concentrate on the rich legacy of Indic religion, so long a hidden resource, on its way to globalization. In a word, we must be wary of such a huge data set.

With our modern perspective we should nonetheless challenge, yet respect the insight of the 'Japerians' and throw a bone to ravenous Christians by citing the passage with which we started:

> In the Western World the philosophy of history was founded in the Christian faith. In a grandiose sequence of works ranging from St. Augustine to Hegel this faith visualised the movement of God through history. God's acts of revelation represent the decisive dividing lines. Thus Hegel could still say: All history goes toward and comes from Christ. The appearance of the Son of God is the axis of world history. Our chronology bears daily witness to this Christian structure of history.

We can process this as follows, with an 'atheist' baseline: we have inherited the 'philosophy of history', found its resolution in the format suggested by Kant, and have 'dabbled' in a Sufistic notion via Bennett of demiurgic powers under translation to a Kantian version, the manifold demiurge. The term 'god' is abstracted beyond existence, the 'son of god' is none other than some earlier intuition of demiurgic power, that is, within existence. The original reference was to the unnamable, IHVH, not 'god'. So we are ahead here. Let us cast this as a new set of metaphysical suspicions: the solution to our riddle is in the biosphere, but with demiurgic powers in nature tending this garden. Note that in our formulation via Bennett the 'biosphere' has 'will' as a cosmic body, and some form of the hyper-consciousness we see in the completion of primitive human consciousness. This is not a life-form with psychological awareness. The terms have transformed. The future of Christianity relative to the modern transition remains unclear.

We can offer this new approach here: the domain of the biosphere, and spiritual powers in nature, in concert. We should point to a *Gaian* hypothesis of the planetary basis of evolution/civilization, in a larger context. The issue of 'god' in history is thus the question of the 'noumenal domain of the 'will' in nature, resolving downward into planetary life, thence history, and we saw its unknown relationship to the question of 'consciousness'. We suggest the 'will's cascade' operates at a far as a lower level, that of the biosphere. We let this pass with a wave of the hand, because it is entirely speculative, but we can see how metaphysical boilerplate is, well, *le plus ça change*. The goal of history is a statement about teleological futures which we have not presumed, remaining with 'directionality'. Jaspers must collate the Axial Age of Israel with the 'axis' of the onset of Christianity. It is a prediction about the future. The future will resolve its 'hypothesis'. This translates the ideas into form an out and out atheist can use, almost...We should leave it at that, mindful we have not produced anything that requires spiritual faith. We are closer to science fiction whose forms all too often become confirmed in practice. But this is not A-life: it is hyper-life. Man already has a primitive footing in this realm, with his 'soul', his potential will and 'superconsciousness'. This account shows how all the myriad confusions of the god beliefs arose.

Our first responsibility is to debrief the Judeo-Christian exploitation of the Israelite perception and mythologization of the Axial Age effect in their isolated corner of Canaanite antiquity. In fact, our larger interpretation is almost more remarkable, a stunning empirical complex, but one whose interpretation still eludes us.

A religious thinker might attempt to apply divinity concepts to our larger spectacle, but the gesture will fail. We have tried to apply an 'evolution formalism' and this was remarkably insightful. There is an irony that as we move toward a post-religious perspective the spectacular character of the phenomenon we have found is almost more a token of a divinity in history: effects across vast distance, remorphing of whole cultures, apparitions of cultural transformation, art, literature in non-random eruption yet all in concert, actions that seem beyond space and time, and 'something' that can scan, intuit and amplify local cultural streams. Despite this theistic interpretations in the conventional sense fail and are falsified by many factors: the action is limited to a precise interval, divinities don't act in cycles. This contradicts a key property of divinity. So what are we dealing with?

> At the risk of confusing the reader let us explore our radical ignorance: it is important to see that the 'middle era' or 'mideonic' religions such as Christianity, Islam, and Mahayana arise after the Axial period *and require a different explanation*. Our model faithfully defaults to the distinction of system action/free action. This does not exclude the action of, viz. 'demiurgic powers' as we have named them as spiritual

powers in nature: the thesis of divine intervention no longer makes full sense. Unfortunately we do not see the real agents of this secondary flowering of the seeds of the Israelite Axial Age, or of Mahayana and Islam, but roughly speaking these religions spring from seeds sown in the Axial interval. So what is 'system action' with reference to the transitions of the Axial period? we must keep in mind that the answer must be a common denominator for five basic transformation zones, with the Maya up in the air: this includes one theistic, one atheistic religion, and a proto-secular phase in Greek, next to the elegant Chinese contrast of Taoism/Confucianism, the later either a variant religion or a proto-secular ideology. It makes no sense to say that a free agent would act this way: the result looks more like a 'biospheric' level 'system' at work, able to act in place and reflecting the cultural specifics in each area of transformation. We will pursue this further later in this chapter: we explore the categories of higher materialism, the confusing interpretations of the terms 'consciousness' and 'will'. There is no contradiction of this higher machine of nature interacting with demiurgic powers of some kind. To make the confusion complete, let us recall that Schopenhauer speaking of the 'will' in nature, referred to something that is similar to 'laws of nature', and that Mr. Bennett took this idea and saw in that the ancient *Samkhya* was saying the same thing in a forgotten language: in short, cosmic bodies have 'will'. We may have found in outline the explanation for the Axial Age, and thence our larger macrosequence. But do we even understand it? We have created some generalized categories that can demand future clarifications. We should be wary of them.

Thus the framework is that of the idea of evolution in a biospheric context, which means, 'brown paper bad holding data showing 'development', on a very high level, and this works because it doesn't chance a hazardous theory, and matches our systems analysis. The surprising bull's eye with such a simple model is very surprising. Our model creates a useful and elegant 'metanarrative' the reader can use to deal with our sudden entry into a stunning hypercomplexity. The model will serve well in this task without resolving all of the mysteries.

The Old Testament account as a 'design argument' for Jehovah will no longer suffice for the modern perspective, but it is still a remarkable classic and the first attempt to observe and chronicle a transition of the type we find in our model! If the modern perspective moves beyond this antique account it is also true that the facts of the case are a challenge to a causal analysis of historical sociology. A design argument ends up hovering over the data all the more once we 'clean up' the text and try to account for the period 900 to 600 BCE.

Conclusion 247

It is good to transit between different transitions zones, from Rome to China, to forestall the obscurity of the Israelite transition. It is a lot simpler than it looks but the overall picture hasn't gelled as yet. We have barely touched the remarkable Chinese Axial phase and its gestation of Taoism and Confucianism, or the equally remarkable Indic and Persian phases, the later being a separate independent emergent monotheism, in the elegant symmetry of Indo-European and Semitic polytheisms shifting into monotheism and amplified in the Axial stream and sequence effect. The blending of the two in the suspiciously convenient Exile is a mystery in itself. A sense of brooding presence broods over the whole spectacle, but it does evoke theism at this point.

The moment skeptical humanism moves beyond simplistic theism a new possibility emerges logically, given the high level and overwhelming sense of design: spiritual powers inside nature, within existence, beings with will but beyond the life realm. Such an entity is also the equivalent to a supermachine or bio-computer operating over a field, science fiction. We have to leave it at that, but a design inference is inevitable here. Is any such factor relevant to our account? Our systems analysis shows that our macro effect is too mechanical, showing an operative functionality that willed agents would not use. But here is the joker: in a Bennett-style *Samkhya*, the biosphere is an 'agent with will' in the larger sense of Schopenhauer. It would also show hyper-consciousness, no doubt, to evolve animals and hominids with 'consciousness'. In a long round trip we re-arrive at the idea of 'Mother Nature', charming. Nature's 'secret plan'. All very Kantian.

But this does not rule out the interaction of both possibilities, like a gardener in a field of nature. And our account, please note, only explains the larger system of transitions, here the Axial Age, and not the outcomes that come later: it does not explain Christianity's onset, for example. The latter is clearly the downfield result of Israelite influences, but the starting point shows its own strange design, and the exact nature of this is unclear. For example the appearance of three founders, in concert, at the outset is highly suggestive. We cannot fully resolve these issues, but have covered all the bases, and protected our novel perspectives against the type of pop theism that so undermined classic Christianity. Our model faithfully reflects the difference in its distinction of system action and free action: the onset of the down field religions is free action after the system transformation of the Axial interval.

Our model explains too many things at one stroke to dismiss, and yet, like a glove, it loosely fits the data, but no more. If we confronted an alien rocket ship systems analysis would tell us what it does and a rough sketch of how its works, but its full data set and actual explanation as a technological artifact might elude us. Just a short survey invokes almost a thousand books: this is Big Data, and it takes off into a larger and larger bibliography. There is still another catch: it is important to consider that our data is fragmentary and we don't know when the macrosequence starts. We can't just attempt to explain the Axial Age, we must find the macro starting point and explain that!

Our account of the Axial Age and the larger system in which it is embedded is thus exploratory, and our use of the term 'evolution' is a formalism of categories, with a macro/micro distinction. This distinction solves so many confusions it must be on the right track. Our model gives us a rough sense of what is going on, and that is remarkable indeed. But the nature of the science needed here is elusive still. The notes will explore new categories, among them the 'hypermechanical'. There is no final difference between the 'mechanical' and the 'hypermechanical', but the range of complexity expands. The mechanical explains known machines, the 'hypermechanical' arises when we try to explain the genesis of art, for example. An example was our idea of the 'causality of freedom' invoked in Kant's essay on history. Kant clearly stumbled into this area in his third *Critique of Aesthetic Judgment*. It is a domain sensed long ago by the expositors of ancient *Samkhya*.

More generally the Axial Age shows system properties, perhaps ultimately some causal system, but on the most intractable and elusive issues: Axial Age Greece is the clearest case with 'output' along the lines of fantastic art sequences, political, economic, aesthetic, philosophic and proto-scientific innovations. These emerge within a given interval and this burst is mostly gone after it concludes, roughly three centuries, leaving the innovations in its wake. This includes emergent democracy and this is a prime case of our hypermechanical: some induction is related to emergent 'freedom' of some kind. An additional complexity arises in the way the macro effect induces creative action as much of the outcome is created under a mysterious influence. This is important to grasp the way the Israelites self-constructed a 'god myth', and this in the context of the macro inducing field we sense but do not understand. We must brace ourselves: this is systems input, output.

The *piece de resistance* is the Israelite: the output is a religion with a new god with the claim that god interacted with the transition, or created it altogether. And this example accurately, though without comprehension,

Conclusion

distinguishes the 'stream and sequence' effect, with its 'stream history' from the primordial up to the onset of the 'age of revelation' a surrogate of our 'axial interval', and its 'sequence history' of the Axial Age transformation, i.e. the Old Testament account of Israel/Judah up to the Exile. How can we explain such a state of affairs even as hypermechanical? Isn't there a design argument here, with a designer? Perhaps, but the designer given by the 'cargo cult' is clearly a fake. We are forced to apply a 'design inference' to the evidence given of a 'fake god'! Indeed. But the whole question here is inscrutable, and a closer look shows the analog isomorphism with the Greek Archaic. It is no accident the Greek Archaic is transparent while the Israelite is intractably obscure. The Greek case makes it obvious what is going on.

> We should conclude with a warning about a dataset so vast we must be wary of conclusions. It then expands further as we embrace the whole macrosequence. We can offer at best a new means of exploration. However, the classic Old Testament interpretation is seen for what is, a mythological account. And a reminder to be wary of this volatile data set. But a good general summary is the model itself that we have created, both crude and effective as a 'metanarrative' sufficiently founded to serve as a guide to further study of what clearly seems to be a teleological system beyond a directional system, one in which our 'free action' can change the result, in theory...we seem to be exiting the last phase of this system as we become aware of it...
>
> First, our metanarrative/analysis has resolved Kant's Challenge, showing how teleology works via directionality, how our 'narrative/non-random pattern' exhibits 'Nature's Secret Plan' and via the 'discrete freedom sequence' (e.g. the double appearance of democracy on the divide threshold) an illustration of the 'progression toward a civil constitution'.
>
> Our metanarrative is that of the creation of a global oikoumene via a system of cyclical directionality whose framework is evolutionary. This system integrates values into the realm of fact and shows its signature in the spectrum of religion, philosophy, science, art, literature and much else. It operates via human free agency which is finally the creator of the larger realization. The macro effect directly prompts the stream of free agency to the idea of freedom which realizes democracy in two (or more) successive transitions. The Sinai narrative passes into something like a Kantian discourse on practical reason, in many ways an historical first. Our system can be seen in the light of two deductions, the first of the appearance of transitions as the logical outcome of evolution turning into history, and the second of the alternation of system action and free action deduced from the hard truth that under-determination

would leave man mechanical while over-determination would co-opt free agency's creation of its own freedom. This metanarrative (to use a term challenged by many critics), in the correct use of the term, passes into the evolutionary biography of a species of hominid entering at the end into the category of *homo sapiens* as an autonomous individuality and civilizationist with a complex instrument of consciousness and a latent 'will' entering into the complex of cosmic life and beyond that the cosmic super-strate that resonates with the newly achieved awareness as species being.

This 'jargon' can be a useful starting point to a succession of trial conclusions. Unfortunately the term 'god' should have graced our paragraph but is exhausted now. Our era will recycle the idea, with or without atheism. But it can be the way to a return to 'direct pointing' we suspect was the original, IHVH. But monotheism has taken up our attention almost to the exclusion of the larger Axial Age with its elegant and profound buddhist world religion, among many other manifestations. This era also gives birth to a new and different post-theological strain, one that will flourish in modernity: the idea of freedom.

> **The Great Freedom Sutra** The modern transition has already stolen a march on the classic yogas of antiquity with its seminal discourses of freedom and autonomy, bursting asunder the spurious authority of the gurus. The passage of free men across the abyss of their freedom might prove not so simple, yet the die is cast, and man is left to the existential reality of his own self-evolution.
>
> **Nature's Secret plan** We have detected 'nature's secret plan' and something far larger in the directionality of civilization: a system in search of the perfect civil constitution
>
> **Revolutionary movements** The fiery radicalism of the ancient 'revolutionary' Christian-type religious movement has shifted in modern times to the 'end of history' revolutions of the modern far left appearing in the nineteenth century.

We should adjourn quickly, our discussion incomplete, to consider the statement of a problem still mysterious and whose solution lies in a larger pattern, and in attempting to understand our own modernity. The material we have found is so rich in content we could reinvent religion on a far superior level than anything in Axial antiquity. But the secular era will claim our attention, and wait on a new form of science. The secular is not anti-religion, and will be the only successor to its era of dominance. The slot for religion so visible in the Axial period is claimed by the revolutionary left. The Israelite Axial phase sent a chase plane in pursuit of the Roman empire, in the Occident, while the same phenomenon claimed the modern revolutionary left, very antagonistic to religion. The future of the modern will be this analogous attempt to mediate the causal stream of modernity with a higher set of values, divorced from archaic religious myths. But we should not forget the humble Münzer with his proto-communist reformation, or the Quakers at the dawn of abolition. Clearly the Axial Age was a set of reformations, as is our modernity. And modernity is obviously seen to seed its own 'Axial Age' or interval in the early modern, this time on the way to a global oikoumene, in the creation of a new set of world views. These were amply summarized in the complex triad of materialism, idealism, and that elusive mystery, the transcendental idealism of Kant, so reminiscent of the realm of the Platonic, and the Upanishads. Religion has yielded to science, but found itself again in the mystery of the modern potential.

We should not despair of our spiritual ignorance or inability to make sense of the 'hypernomic' realms. In one century since Jaspers codified the data of the Axial Age we have made more progress than in the previous millennia. But the Kantian critique of metaphysics restrains us, so far. We have in any case a rich nature of history that unifies the sacred and the secular, and leaves us in the wake of the 'modern' 'Axial Age', ready for a new exploration of the scale of nature in the evolutionary emergence of last and first men.

Notes

We have left the reader with the pieces of a larger puzzle, and a default 'systems' model that can roughly express the narrative emerging here. In this section we will move between 'system accounts' and their possible source in a biospheric or *Gaian* system, with the issue of 'design' factors left as questions, with in addition a new form of hypothesis based on the ancient *Samkhya*, none other than our hypermechanical.

> We should reissue a new terminology, which skeptics may freely protest, to replace that of 'demiurgic powers' as corrupt Sufistic or demonic concoctions: the source idea is Platonic in the singular: the 'demiurge'. The singular/plural ambiguity is apt, but we must be wary of what we mean. Let us designate the 'manifold demiurge', or 'demiurgic

Brahman/Atman', or 'SPR-MAT-X factor', as a category of entities within existence as beings relative to *Samkhya* guna levels 3, 6, 12, i.e. entities with 'will' and 'consciousness', active within the scale of a solar system, distinct from 'alien life' (which is 'alive'). Our biospheric system is 'hypermechanical' and is too mechanical indeed to follow a design argument, but we can't rule out hypernomic entities that are active in various phases of cosmic life. The Christian idea of the Heavenly Host might help make the point (a term we don't use and didn't define). The onset of mideonic religions require a new form of explanation and our phantom neologisms are potential candidates.

The hyponomic, autonomic, and then the hypernomic are a new division of material reality: the hypermechanical belongs to the hypernomic, and we suspect that 'consciousness', for reasons that must be carefully examined, is man's bare first taste of this higher realm. Isn't this really the same for animals? Perhaps the hypernomic pervades the autonomic and the bare vital awareness of the animal realm anticipates man's first steps into this realm. In any case, we might take seriously the idea of 'consciousness' as a 'cosmic' or hypernomic energy.

> We are using an idea or framework of ancient Samkhya, via an update by Bennett. The Gnostic background he emerged from will attempt predatory and proprietary ownership of Bennett's thinking. We have taken nothing from the latter save the one clear instance where that is not true. Sufis have no claim on such an ancient aspect of primordial *Shaivism*: *Samkhya* goes back thousands of years before figures such as Gurdjieff, who muddled everything he touched.

We must depart from Bennett's formulation to reset the issues as abstractions in the realm of possibility, but subject to the limits of the metaphysical. But we have navigated the dangerous terrain of Bennett's Gnostic confreres almost as a warning about dangerous unknowns with demonic interpretations. And we have left the barest whisper of a hint about the dangers of the unknown legacy of sacrificial religion which infested primitive men, with a question about a misfortune that might have befallen the Axial synchronous domain of the Maya.

Our remarks about 'demiurgic powers' from Bennett are the same old speculative hand-me-down from antiquity, this one from Sufis. But Bennett recast them and then threw them away on a Gnostic 'food for god' equivalent that is entirely without sound legitimation. We throw the terms back into the witch's 'stir', and exit via the entry point: Kant's speculative notion of 'the demiurge'. But such entities have the same problems as yogis, in the 'non-dual' cosmic reality as atman as Brahman, a warning such entities are

Conclusion

even less intelligible than 'enlightened Buddhas'.

We must without delay create our own version, to distance ourselves from the Gnostic exploitations of spiritual hierarchy. Our thinking has already shifted to a better understanding, so we are free of the false framework Bennett inherited from shark sufis. We have to issue a warning to be wary of any such notions from known spiritual con-men. Sure enough, Bennett attempts to connect his idea with an ancient and unattested legacy of sacrifice, man as 'food for god'. We must reiterate aggressively that the Axial Age began, and Christianity among other movements, led world culture beyond the terrible legacy of animal and human sacrifice. We must be wary of attempts to revive them in occult modernity: here Bennett was naive in his thinking and must restate the simple idea in a neutral form: the debates over god are metaphysical and refer to something beyond existence. There is a purely logical possibility of entities of higher 'material spiritual' definition existing within nature, existent, but beyond the autonomic or 'life' realm. This corresponds to the possibility of beings with 'will' and 'consciousness' (a term requiring careful use) without bodies of the evolutionary/biological type. The logical possibility is there and strongly suspected.

> We should answer the protest of theists that an operation on this scale shows all the evidence of theistic action in history claimed by the Old Testament. We have noted that any explanation must be about effects within existence, and that 'god' is beyond existence, and equivalent to a concept of the 'void' producing something from nothing. The paradoxical reckoning of the 'edge of space' paradox shows that the account is antinomial in a Kantian sense, the relation between 'existence' and 'being beyond existence' remaining a metaphysics we have not mastered. And it makes little sense to claim a 'god' beyond existence would operate only in cycles, and produce an atheistic and theistic religion simultaneously. This is really something that triggers the cultural realization of potential in the culture zones in place.

Again, the only real issue here, to make a hasty departure from Bennett's 'demiurgic powers' is that there exists a category of possibility of 'spiritual/material existence within nature', within existence, that is not 'supernatural' or 'god'. Since it is not autonomic, that is, 'alive', the only two categories we have left are 'consciousness' and 'will'. Entities within existence that have will and consciousness beyond life foot the bill, although we know nothing of such things. But they are candidates for 'design arguments' about history, if only to forestall creationist obsessions so rife in the Old Testament. There is a possible rubric for a science of the future here. We discard the term 'demiurgic powers' and consider the issues strictly as abstractions. The idea in fact pervades monotheism in the decayed versions known in the crypto-pagan mythology of 'angels', beings with consciousness and will but no bodies. The correlation is an invitation to confusion. Best avoid it and start from scratch.

But, to reiterate, Bennett's updated version of demiurgic powers is compelling, yet compromised by occult confusions, but he introduced a possibility neither theists nor atheists thought of: spiritual powers in nature, beings within existence, yet beyond life, with 'consciousness' and 'will'. There is every possibility this is the case, and speaks to the possibilities soon to be foreseen in the progression of advanced technology, which will include the quite different but related idea of alien life. But the latter is 'alive', evidently and very different. The idea of 'demiurgic powers' is still archaic and mixed with ideas of sacrifice, a wretched botch on Bennett's part. But it leaves the question of what was going on in so many millennia past with the issue of sacrificial primitive religion, practices whose barbarous history is not properly documented, but which clearly persisted until the era following the Axial Age. And we see the isolated case of the Maya, whose status in the Axial spectrum is unknown, deviate into an extravagant sacrificial religion, our warnings being more than sound. We can recall our discussion in the *Preface* of the ambiguous and alarming case of the Maya civilization. If we study our discrete-continuous model we can see how lateral relative starts can initiative advanced civilization in cultures that have missed the earlier parts of the macrosequence. Bennett thus may be right, but he confused two things.

We have simply deleted the concept of demiurgic powers with a replacement, entities boringly tokened 'SPR-MAT-X', i.e. 'spiritual entities' in the realm of nature with consciousness and will, but not part of the life process. This is a distant analog to 'hypermachines' of supercomputers, which are not in the 'life realm', to make the point clear. Many traditions have attempted and failed to consider such questions, with demonic confusion in the wake. Our approach grants nothing but restates the issue: it is entirely within the realm of the possible for there to be a higher materiality with entities showing 'will' beyond the autonomic. Again, *Samkhya* was always warning that the spiritual was really 'higher material' and wasn't the real spiritual. In our language, the 'spiritual' is the 'noumenal analog' to the phenomenal hypernomic, like the god realm denounced by Gautama.

However, since we have no direct evidence of this kind of entity, we are done, but the potential of this kind of explanation remains for the future,

Conclusion

and help to sort out the extravagant cargo cult of the Israelites. But at times men have no doubt felt the influence of such beings, and this is part of the confusion between the Axial Age, or short interval, and the 'age' that follows. The religions like Christianity emerge after the Axial Age, we must not forget, and require a different explanation. The Axial Age is too vast even for a demiurge: it smacks of hyper-mechanics, our 'systems analysis', and operates over a range that is stupendous, on a *Gaian* scale, manipulating time-slices of cultures in short bursts over tens of millennia. Spiritual design agents would not operate this way, multitasking over separate regions, and then so promptly switching off after short impetus. It smacks of high technology, not divinities. Science is on the threshold here, we hope.

> It was Schopenhauer who stumbled on this new understanding of the 'will in nature', but his 'will' as 'thing in itself' is perhaps conjectural, and we would have to modify his explanation: as did Bennett in the process confusing the phenomenal and noumenal. The latter mistake is the escape clause from his concealed mix of religious and secular concoction.

In a word we confront a new type of problem for which have only the most rudimentary tools. The point is that we have not invoked the supernatural with this, the conception being a variant somewhere between crackpot 'angelic mythology' and questions/theories about alien life. This is not atheism, but the perception of secular humanism that pop theism is an idolatry we must avoid. Speculative science fiction often predicts right, so we can leave this tabled in our Axial theory toolkit. This is not as such atheism: the concept (sic) of 'god' is of something beyond existence, and metaphysically unknown, although in some antinomial sense present to all space/time, with pleas to string theorists to resolve cosmophysics. The distinction of being beyond space and time is logically valid, its reality very hard to really understand. Kant warns us to be wary of metaphysics.

> To this we should add the yogic 'discovery' that consciousness is or has an octave of self-consciousness, perhaps beyond mind or body, a thesis hard to verify, and as hard to understand, with still another aspect as 'enlightenment'. In any case the Axial Age is too mechanical for this other type of speculation about the demiurgic manifold and falls into place as a kind of supermechanics of nature in a *Gaian* field. This *Gaian* field looks inscrutable, but shows in reality a simple trick: each culture in the Axial shows creative transformation carried out by the people in place: the mysterious larger field 'does nothing' beyond injecting a kind of creative energy or potential.

The question of consciousness was always the hard problem, and is here scrambled into several 'confusions', and now we see it may be still harder than we thought. It was first named by a westerner using the phrase 'cosmic consciousness'. Bennett thought 'consciousness', more like what we mean by 'mindfulness consciousness', as the lowest level of cosmic energy in a tetrad of increasing abstraction, and that bodies of consciousness were the outer form of demiurgic mysteries. In various *Advaita* traditions, instead of seeing consciousness as a 'state' outcome in the evolution of animals, it is a pervasive substrate that emerges in different forms in the animal realm.

We inherit one last confusion: the status of 'enlightened buddhas': this mystery has its own lore. 'Enlightenment' transcends the 'god realm', thence demiurgic powers, would not be a 'state of consciousness', and would renounce the category of will. The term 'consciousness' has suffered wreckage: we should note the alternate usage of classic *Advaita* with 'consciousness' as the groundstate of the cosmos, and what is left after all the negations of existing bodies strip away illusory 'selves'.

We almost have the plot for a good Hollywood movie, and are drawn beyond the solid gains of our model into speculative possibilities. If these are even remotely close to being the case, we have a clue to a new science of the future. What do we mean by consciousness? The term is shifting between different discussions unchanged. We can't complete our argument unless we can understand this. It is important to see that 'consciousness' in *Samkhya* (whatever we concluded given translations of such terms) is external to the cascade of gunas. Maybe that is confused, but it has a deep insight. It is therefore remarkable that the Axial interval in India includes the remarkable distillation called *Advaita*, along with the Jain *teertanker* sequence, and then the buddhist send-off into globalization just after the 'divide', around 600 BCE. The *Advaita* has a better canonical rendering of the meaning of 'consciousness' and would require a revision of any view of 'evolution'. It might be that man, or animals, begin to exteriorize phenomenal versions of the greater 'consciousness' which is a cosmic energy (or pace *Advaita* the cosmic reality itself). These are classic results injected into world culture in the Axial period, so our usage is ironic.

This might actually be simpler and more elegant than trying to explain how 'man', or the 'animal', 'evolved consciousness' as a property of organismic neurology, but a very heretical notion for current science. Perhaps both views are correct, requiring greater care in our use of terms like 'consciousness'. But it seems to imply that primitive hominids are emergent Zen buddhists. Well, OK. Again we see that an Axial Age production is still relevant to our

Conclusion

modern perspective. In fact the question of Indian religion is one of relative transformation in the Axial Age and its 'stream' aspect goes back many centuries or millennia to the dawn of civilization itself. The Axial phase 'modernizes' the ancient cult of Shiva, the final source of the later spiritual psychologies such as the buddhist and the '*Advaitist*'. It is very hard to get this earlier history clear, since it has been scrambled with the Indo-European heritage which was introduced at about the same time the 'aryans' entered Greece in the second millennium BCE. It could be that the Indic legacy goes all the way back to the Neolithic, or to at least the era after 3000 or so BCE.

> One of the mysteries of the macrosequence is the source of the primordial *Shaivism* that produces yoga, tantra, and the Jain predecessors to the Buddhists. There is a perennial quarrel here between yogis and sufis, but these play in different fields, the paths of will versus being. The Indic stream may arise in the Neolithic in some primordial Axial Age of an earlier period, as considered by the philosopher Danielou. The source precedes the Aryan entry into India.

The sophistication of this legacy is such that we are left with a question so far unanswered. The Axial phase does what it always does, picks up stream elements and recycles them as 'sequence output', between the early Upanishads (carrying *Advaita*) and the final 'Buddhism' which, in exact concert, with post-Exilic Israelitism, starts to globalize into a world religion.

> As for the design arguments in the Old Testament, it may be that some designer, maybe a demiurgic power, created a myth of a fake designer, designer, in compassionate predestigation for Hebrew primitives. But there we go again, with Bennett's ambiguous spiritual powers. This is the terrain of spiritual illusions, so we should be wary. But our model suggests a more obvious answer: a primitive tribe began to notice the Axial transformation and began to weave a spider's web of theistic mythology. Their religion is thus their own realization in a field of Axial Age effects. It is not easy to explain what they experienced in the language of historical sociology. The denizens of the modern Iron Cage will be lead away whimpering, perhaps in straightjackets. The situation is worthy of H. P. Lovecraft.

That fake design may generate still another design inference is an almost inevitable explanatory calamity, and perhaps gallows humor. In the reckoning, monotheism was a lot of trouble. Look at simple Buddhism plain, elegant, and practical, with no metaphysical nonsense. But there was a catch: is yoga at any time the same as Buddhism generated in the Axial mechanism? If not, if Gautama is a creature of the Axial meta-system, the mechanics will catch up with his beautiful product in the end, enlightenment should be beyond such a metasystem. Another case for the *Samkhya* hypermechnical! Gautama performed his task: he left the question raised by the forest yogis to the greater humanity beyond India, a true Axial Age production. The point is to ask if 'enlightenment' is part of the temporal stream. It is a remarkable question: can 'enlightenment' be system generated. Best to file it away for the nonce.

As we have seen the factor of free agency provokes the extreme non-linearity of a hybrid mechanical/willed systematics and it is much simpler to exit such analysis and do what scientists do in practice: chronicle an empirical sequence that reflects the dynamic in question. But in principle our explanation seems right because it is about the 'evolution of freedom' and we see the result in progress as man alternates between induced and self-realizing freedoms. But in a nutshell we have used system analysis to conceive a new form of mechanics, something intuited by Kant in his third mysterious critique, of aesthetic judgment. Our system shows so much art in motion as cultural dynamism that we must hope Kant with this gesture saw the future of the sciences, now stuck in scientism, the dreary cult of scientists frozen in the Iron Cage.

Our discussion of an evolutionary macrosequence, taken empirically performs the task of explanation prior to theory, as a visualized model, and this result is a clear metanarrative or 'book with chapters', each chapter an epochal age-period, starting we Egypt/Sumer, then the Axial era, and then modernity. Since we deal in 'relative beginnings' we can start anywhere, and annex a larger history with new chapters from earlier eras, such as the Neolithic, which we suspect contains two earlier stages of our account. The idea of a sequence was complicated by the ability of this majestic system to split into sidewinders, and this phenomenon we see with stunning impact in the Axial Age's multipole synchronous phases. The larger system idles from its focus on directionality as the splitting effect chaotifies direction in the intense dwelling on local places. The two great religions seem to wish to usurp direction, the future here, but our macro system with a drastic frontier effect switching to far Western Eurasia, the Euro-field, resets direction with the dramatic impact of modernity as a successor transition.

These remarks are a reminder of the complexity and vastness of our material and we need to retreat to the 'bird's eye view' of a simple narrative, or 'metanarrative', a term criticized by postmodern critics, one that can assist in digesting a discourse so colossal we will end by evolving larger brains to understand it. Our model will serve this purpose very well and 'reduce'

Conclusion 259

the complexity to a rough chronicle ending in the onset of modernity. We must in the end find the 'answers' in the modern 'axial interval', if there are any, and there we see the dramatic onset of the 'secular' recursion of the Reformation followed by a rapid dismantling of Axial Age religions, with are passing into the obscurity of past history the way Egyptian religion did in the wake of the Axial Age.

Modernity seems to suggest the 'secular' is the antithesis of 'religion', but this is misleading. We cannot banish discussions of consciousness in the name of the secular. We may invent new categories beyond that of the now archaizing 'religion', but the core of the sacred is amply present in the rich symphony of effects we see in the modern transition. We should leave it at that, for the nonce, with a model of the Axial Age that shows a richness of content almost overwhelming in its depth. It wouldn't be hard to create a religion of revelation based on the modern transition, but we should do better to reinvent our whole understanding mindful that 'modernity' is supercomplex where the Axial Age was complex, and can't be reduced to a canon of religious gibberish.

The analog track for religion in the wake of the modern transitions strips out all religious concepts in the period of Feuerbach and proceeds along a revolutionary path toward the mideonic reconstruction of social culture via a universalizing ideology, none other than the realm of socialism and communism. These have usurped the similar action of religions in antiquity. The explosive New Age movement seems intent on recursions of religion in the modern period, but we have to wonder if they aren't 'running on empty' as doomed recursions of an ancient impulse.

We should consider that most religions are confused muddles of the transcendental idealism of figures like Kant, and more clearly Schopenhauer. The modern period stole bases almost immediately and this will force a total recompute of the whole axiom set lurking in archaic religion. The real question is the ethical will, self-conscious realization, and temporal/transtemporal realization natural to *homo sapiens*, forever lost, and found again in the modern world attempting to grasp the nature of evolution in the progression of hominids into the species 'man'.

4. APPENDIX 1: AN IDEA FROM SAMKHYA

Our exploration of the Axial Age could use one more interpretation, one already indicated in the *Preface*, and in the previous paragraphs, using an idea from the ancient *Samkhya*. We have also used and replaced the concept of 'demiurgic powers' with new ones requiring redefinition: e.g. the demiurgic manifold, demiurgic atman/brahman, SPR-MAT-X, etc... Metaphysical unknowns are dangerous, like bad pointers in C programming. Their value can be anything and demonic anything might well adopt the term. We are talking however about a purely theoretical possibility that deserves discussion.

In this variant of *Samkhya* we find, in our distinction of hyponomic, autonomic, and hypernomic realms, the higher materiality of 'spiritual nature' beyond life, which can have properties related to 'consciousness' and 'will', in forms unknown to us. Note the distinction of consciousness and life as vital awareness. We don't know precisely what 'consciousness'

Appendix

means beyond the simple 'awareness' as an organism nature gives to us, but the affinity of man to meditative quests shows he has a sense, and often some 'conscious' state, of this deeper reality. First *homo erectus*, and then especially *homo sapiens* began to 'bump' into this mysterious hypernomic milieu, reflecting in a frequent double terminology the experience of 'self-consciousness' beyond 'consciousness', and states beyond that. The confusions of 'consciousness', 'will', life/vital awareness, spiritual powers in nature and beyond nature, explains at once the hopeless confusion of ancient theologies as they emerged from the Axial period.

> We have introduced the idea of 'demiurgic powers' from the thinker Bennett. In what follows we have replaced this term with a neutral 'SPR-MAT-X', in order to evade proprietary usage from Gnostic operators who will claim to mediate such entities. We have no knowledge of such entities, but we need in principle to assist the debriefing of monotheism and its many abuses as pop religion by considering the logically counterpoint idea of entities with 'consciousness' of some kind or will but beyond the life domain. Such entities are distinct from the hypothesis of a transcendent 'god' logically confounded by antinomial contradictions as beyond existence and space-time. It is small wonder modern atheism attempts to sweep away this morass. But the domain of SPR-MAT-X entities points to the logical slot in '*Samkhya*' of 'higher spiritual nature'. The data of the Axial Age is beyond all of this: it is too complex even for simplistic design arguments.

The Israelites detected the macro effect, a stunning achievement, but in the process confused the clarity of the history they lived. We end up with the Greek Axial interval as the key to what we see. Monotheism emerged as 'output' of the system in our sense, and yet was applied to the mystery they detected. The Old Testament collapses (unless it is a sci-fi-like account of a designer promoting a myth of a god/designer to a primitive people). We can forgive them their confusion: we have no science that can explain, e.g. aesthetic effects, although Kant stumbled on a first intuition here. Our larger pattern is so spectacular we could be struck dumb and reinvent a new religious historicism ourselves. But the Greek Axial Age gives us a clearer portrait of what is going on. Let us recall our suspicion the original 'monotheism' did not speak the forbidden term 'god', instead pointing to the 'unnamed' IHVH. We can certainly appreciate the point and must refrain from using pop theism in our account.

We can see the possibility of a new range of forms of explanation: the clear perception of a dynamical system using an 'evolution' formalism, which does not exclude design arguments. The latter is a 'brown paper bag' to hold the data showing a discrete progression. These new frameworks of explanation include: new concepts of 'unknowable' divinity based on Kantian noumenal boundaries to the phenomenal, in the 'edge of space' antinomial superset of 'existence'. God, or the Void, is beyond existence. Here theistic historicism is replaced with design arguments displaced to 'spiritual powers within nature', within existence. We can offer the speculation that a clear perception of the *Gaian* scale of our phenomenon of the Axial Age suggests a biospheric level induction of evolution. We have tried to extend our totally inadequate language to distinguish new concepts distinguishing life, consciousness, and 'will'.

The elements of Schopenhauer and J. G. Bennett have left a question mark about 'demiurgic powers' in nature which have 'will' and 'consciousness' but which are not 'alive'. Related to this is the question of space-time itself and the way in which 'consciousness' and 'soul' as an aspect of 'mind' are ambiguous in their relationship to standard geometries. We are certainly not quite ready for this level of complexity! The realm of modern 'spooky physics' provokes precisely this kind of question of effects that are trans-spatial.

> An irony: the modern 'secularist', debunking the Old Testament, will end up inheriting the whole account and savor its primitive detection of the macro effect.

But the two thinkers are not quite the same and a new hopeless complexity confounds the discussion with the ambiguity of noumenal and phenomenal aspects of 'will'. This was also present in the triple distinction of hyponomic, autonomic, and hypernomic realms. The hypernomic realm is precisely the realm explored by the ancient *Samkhya*. This is the phenomenal still. The noumenal corresponds to something beyond this, and beyond knowledge: the point is that the hypernomic is still within materiality, and should be open to sciences of the future. We are used to thinking that entities that are alive are conscious, but we have suggested via Bennett that 'consciousness' enters life, but is beyond it: consciousness transforms to a higher octave as 'self-consciousness' which is beyond the organism, very controversial, and not as yet scientific. But the distinction, if it could be understood, clarifies many things. Ancient theists, in blessed muddle, clearly made this distinction in garbled form: angelic forms, soon the object of a new neo-pagan mythology, had consciousness, and will, but were not living beings living in bodies of the type we know on the biosphere of our planet. We have stumbled into *Samkhya*:

> **An Idea From** *Samkhya* We have no conceptual language to deal

with our larger 'macro' effect. We must distinguish that effect from the genesis of religions in the middle periods of our pattern. We see a remarkable pattern of dynamism, and yet we can't rule out design arguments. We introduce an idea from *Samkhya* to distinguish the 'material' aspect of the 'spiritual' from the noumenal unknown which is the antinomial omnipresence which incoherent theism has considered 'god'. To consider the point take the 'antinomy' discovered by Plato of the 'edge of space': at the edge of space can we reach beyond it? Here three different theologies arise in splendid unity and contradiction: the idea of current physics of 'something from nothing', the buddhist idea of the void, and the creationist idea, suitably upgraded beyond personalized theism, of a supreme god beyond existence, IHVH, nameless and conceptless. All three of these are really the same (and equally incoherent of antinomial).

As we complete our survey we see that the significance of the Axial Age is that of its place in a progression of epochs in world history. We called this the 'eonic' or 'macro' effect, to replace the term 'Axial Age'. The result is an incomplete picture, but as such so spectacular that we can appreciate the Old Testament's sense of detecting a higher power in history. Small wonder the Israelites created a religion of history. Our humanist critique fails to see the way the 'Israelites' detected a transition in motion. It is not surprising they considered processes involving cultural regions over time could only be the 'acts' of a divinity.

A Higher Power Acting Through History It is almost egregious to throw our data into the grabbag of 'self-organization'. The eonic effect fills us with a sense of an almost ominous presence, of a mysterious process or action operating throughout history as a higher power. We see fine-tuning down to the level of poetic meters and even the whole genre Greek tragedy that might leave us floundering in design arguments. We need to realize that divinity would not act in this way. Conventional theism/atheism will not help us understand this situation.

In fact we have rediscovered, perhaps, the elemental sense of universal history first intuited by the Israelites, pointing beyond god idols to IHVH, before that degenerated into monotheism. We have lost that tradition, and need to steer well clear of it. We cannot under any circumstance bring 'god ideas' to our depiction, at the risk of corrupting our clarity with the confusions of false design arguments. That would truly wreck our account. The same can be said of the sterile atheism based on the metaphysics of Darwinian natural selection. The depiction of 'evolution' using systems analysis keeps our account honest.

The term 'higher power' has no intrinsic semantic reference to 'god'...

However, the signature of systematics creeps up on theology. The design sense remains strong, but our model exposes the idea of 'god' in history as falsified (in most senses, but says nothing about the noumenal/antinomial 'edge of space' omnipresence). The sense of design is a teleological system in motion. But designers might lurk in relation to such a system. The modern transition shows how the cases in the Axial Age work. There we find no trace of a theistic design argument, but the data does not correspond to reductionist historicism either.

> We have not stripped 'god' out of the age of revelation and promoted a purely mechanical substitute. Our model is not a theory. You can, given the stunning data we have found, certainly point to 'seeing god in history'. but the problem is that you must define/redefine 'god' first, and create a semantic community that shares that term in that sense. This is obviously impossible, now, but the Israelite impetus began the classic trial in this vein. Monotheism emerged to eliminate polytheistic confusion, but ended by creating a truly confused semantic universe around the term 'god' in many languages. Let us note the curious remnant notion of the 'unspoken name of God', IHVH. Something has been lost and we have the suspicion that the legacy of pop theism of 'Jehovah' is a distortion or decline, an idolatrous scandal we must pass by in compassionate anthropological hauteur. We must be clear how misleading the Old Testament mostly probably is. The crystallization that took place around the divide era (600 BCE) has lost the real history that led up to it. And the spectacular manifestation of 'prophets' is all the more spectacular and mysterious. Try this as an exercise: consider the modern 'divide' and study Kant, Hegel, Schopenhauer, these cap three centuries of the Reformation, and in many ways displace it. Then a generation later we find Feuerbach. And this reorients a whole century of modernism. Now consider our perspective two millennia from, if we barely had writing, or any documentation of modern history. We would be hard pressed to make sense of this situation and might conclude that the modern transition was about the views of Feuerbach, and forget there was a Reformation. In fact, this has already happened. So what the real history of 'Israel/Judah' was remains ambiguous. The hidden 'reformation', still close to the general components of Canaanite 'religion' with Egyptian/Sumerian direct descendants is not clear to us.

There is a long range effect that dwarfs the Axial Age. And the historicism of 'freedom' is a far more intelligible interpretation, and this was the 'sleeper' stream running through the Greek Axial. Its quiet hyper-naturalism is far better than a metanarrative of the supernatural because its factual basis is documented by the history of modernity, and yet, as Kant has shown, the idea has immense depth in the counterpoint to 'Newtonian' dynamics.

> We should be wary of the immense complexity of the spectrum of general outcomes, Confucian, Indic/Buddhist, monotheistic, Greek general spectrum….and the Mayan exceptional case…
>
> The Persian/'Israelite' *tour de force* is a complex generator/experiment moving to clear away polytheism, 'idolatry', animal sacrifice and create a new type of oikoumene/civilization.
>
> India shows complex a Axial transition in three visible aspects: the Upanishadic collation, the Jain sequence of Teertankers leading to Mahavir, the onset of a globalization vehicle in buddhism in exact concert with the Israelite.
>
> China shows a dual parallel birth of Taoism and Confucianism, and then a later 'mideonic' blend creating the classic 'Zen' legacies.
>
> The Mayan case is still hard to understand but seems to be a cycle behind, and ritualizes an archaic 'sacrificial' legacy that will prove dangerous in the end phase of the Aztecs…We are not sure we should include this case, but it fits the general rubric of stream and sequence and 'Axiality', and it demonstrates a real global mystery…Isolated civilizations are at risk: the Old World cases all show a manifold diffusions and help each other out.
>
> The Greek case (with its variant Roman) shows a final flowering of aesthetic polytheism (cf. Greek tragedy), but is moving in a different direction, our 'secular', with the birth of science, a philosophic revolution, republicanism/democracy, a 'lyric' age (collation of epics, followed by…), and much else…
>
> These legacies will spawn 'middle phase' religions such as Mahayana, diverse Chinese outcomes, Christianity, Islam, and 'Judaism' as such. Note that will the next transition will focus on the Euro-zone, there are numerous isolated 'reformations', e.g. the Sikh hybrid in the early modern….It is important to see that the dynamic of these religions, although seeded in the Axial period, is different, requiring different explanations. Note that Greek/Roman polytheism, and 'Israelitism' fail to transcend 'animal sacrifice', with disastrous results, arriving only with the later religions….
>
> Our "stream and sequence' analysis fits all these cases to a T, viz. the 'streams' of Indic, Canaanite/Abrahamic (?) legacies and the Axial interval transformations to proto-monotheism, and buddhism…

We need to be wary of the stunning macro pattern we have uncovered: its outcome is modernity itself, whose interpretation shows something still more spectacular than the Axial Age. The secular, so clearly prophesied in Axial Greece, has taken the stage as the world system moves in a new direction. This account does not preclude design arguments, but the classic Old Testament account is clearly a construct of mythology built around a transition interval (with a stream to sequence lead up). We have outlined this situation with distinction of theistic powers outside of existence, material 'spirits' inside existence, with reference to the 'demiurge' of Kant (Plato). Atheism in its modern form does not grasp the possibility of spiritual powers in nature, inside existence, 'god' by definition beyond existence, and thus incomprehensible. The Israelites sensed this and clearly distinguished IHVH and the '*elohim*'.

Divinities don't act in cycles. But the hyper-machine in question shows elusive properties including the aesthetic. Its design aspect is overwhelming but this implies nothing about 'intelligent designers'. We introduce the term 'hypermechanical' below, to clarity the hopeless confusion here (we have considered distinctions of hypo-, auto-, and hypernomic levels. The 'material/spiritual' pertains to the hypernomic, by hypothesis. The 'macro' effect begins to provide suspicious evidence of the hypermechanical.

> The simplest snapshot of the larger context of the Axial Age is that of a discrete/continuous model operating in a 2400 year frequency cycle. The success of such a crude model is striking. We have discontinuous intervals (in a sequence) inside continuous streams, and these are short three century bursts with divide lines at the end. We see only three beats in this series, but the continuation is clear enough for a hypothesis. This mainline of this directional sequence (teleological) would bypass almost the whole planet, so we see, beside the 'sequence', sudden splitting lines in parallel. These intervals spawn diffusion zones, or new civilizations/layers. This minimaxing directionality shows no distinction of sacred and secular. This argument does not rule out 'designers' tending such a machine: the doubloon aspect of system/design is strong. The macro effect implies properties we don't understand, such as scanning regions. The larger picture shows a mysterious global/Gaian process creating a garland of flowering localities across Eurasia and beyond. The later religions are a different process and spring from seeds in the transition but enter the phases of free action, and that could include men, hidden spiritual traditions, or unknown 'spiritual powers'.

Appendix

We have introduced the idea on the sidelines of 'spiritual/material powers in nature', viz. to replace the term 'demiurgic powers', 'SPR-MAT X entities', within existence, defining 'god' as beyond existence, a point of clarity lost in centuries of theology. 'God' beyond existence resembles the 'noumenal' and is beyond knowledge, leaving atheism/theism in stalemate.

There is nothing contradictory (supernatural) about entities inside existence and beyond man. The 'mechanics' must be specified, perhaps by a science beyond man's capacity at this stage of evolution...

Such entities could be alien analog 'life forms' (unknown on the planetary system of man), or they could be beyond life in the realm of the hyper-mechanical...The element 'consciousness' is one possibility often proposed for this category, in a different sense of the term 'consciousness'. Consciousness could be a ground state for a higher level of hypermechanical states of existence.

We can specify science fiction elements such as

beings inside existence with 'bodies of light or consciousness'

electromagnetic fields able to apply computational powers at the mental level over large regions...

We can dismiss these speculations, but we see the evidence of supercomplex mechanicity in the Axial Age...

Our dubious statements about demiurgic powers will confuse the issue, and should be dumped forthwith. But they were introduced with a warning (cf. the *Preface*). We will use only semantic entities we control, and discard them at once if necessary. But the narrative of 'demiurgic powers' of Bennett remains of interest via the history of sufism.

 Such notions can be pernicious. Demiurgic powers would mediate via human vehicles and sufi sharks would volunteer at once for a 'piece of the action'. But these can also explore logical alternates, and we should air them and let them play out. Let us note the confusion that arises with monotheism, all the other transitions of the Axial interval are much clearer. That is a sign the Old Testament is metaphysically incoherent. We must set it aside and compare the Axial interval to the case of Axial Greece, as an analog. The Old Testament is not a usable record of the Axial interval of Israel/Judah. It is the output of the transformation!

We have discussed the modern transition because it shows a deep design but no signs whatever of designers. The record is 'clean', and Spinoza appears in the early modern with grim determination. It is important not to give ambitious spiritual politicians any angle on the supernatural. They will soon be their representatives on earth, but we have picked one example, 'demiurgic' powers as an example. Let us list the possibilities:

> the transition: this is hypermechanical, and probably locked from outside designers. This hyper-machine operates on a scale of ten thousand years or more: designers, demiurgic powers, might initiate this machine, but wouldn't intervene in its macrosequence without good reason. The Israelite account is thus mythical, but... The whole notion of designers fails, for the evolution of civilization is bound up in a larger systematics that goes back to the start, a Goldilocks principle... Confusing teleology with designers is very easy. But the 'designers' of such a machine would surely monitor its progressions.
>
> the post-transition: we can't rule out the possibility that designers take up the phase of 'free action' and assist the realizations of the transition. The quick start religions like Christianity and Islam might be related to this, but as realization of the potential seeded in the transition. But Christianity looks far more like a later realization of the inner potential of the transition, correcting its crudities.
>
> The overall system shows both factors, a directional system of hypermechanical effect and SPR-MAT X factors either connected, or associated with the larger 'garden' of nature.
>
> These religions show a 'goal' of moving beyond polytheism and paganism, but the result via free agency is strangely 'incompetent': the distinction of system action and free agency, perhaps.
>
> The legacy of buddhism shows far less of the strange confusions of monotheism...The same is true of the Greek Axial transition. The latter is the only serious data set we have of the Axial Age. But 'Israelite'/Persian transitions set in motion long range culture creators called 'religions' which attempt to replace paganism with monotheism, oikoumene generators that create new civilizations as overlays.

This might explain the fact that 'religions' like Christianity generate illusions of revelation but are outside the transition. Spiritual entities could 'read' the transition and initiate downfield religions. The accounts of Jehovah in the Axial transition are completely misleading and later interpolations. But the Israelites did detect a transition, which is remarkable.

> ...what do we mean by hypermechanical? We should consider three levels; the level of physics, the hyper-mechanical biological, and a hyper-mechanical X at an unknown higher level, the old fashioned 'spiritual', the higher *gunas* in *Samkhya*, level 12, 6, 3...Such notions will get in trouble! they are not science. But we see the high level hypermechanical

in all our transitions, which can process ultra complex objects like 'culture', literature, religion, etc... Hopeless confusion has arisen in taking distinct things as 'god'. *Samkhya* had this idea: the *'prakriti'* (vs. *'purusha'*) will fool you, and goes very high, but is still 'material', or phenomenal.

This overall data set is a classic 'four-dimensional black box' whose interpretations stretch between old fashioned design arguments and our Kantian resolution of the dynamics of freedom, with our systems model. The model is the correct way to strip out false design fantasies. A 'model' is not an explanation or causal reduction. If you encounter a space ship of aliens, systems analysis will describe what it does, but not produce a causal explanation. And all explanations must reckon with 'free agency' inside 'system action': fallible men with superstitious beliefs saw a transition unfolding and thought it the action of a divinity, thence 'the' divinity. In the midst of this their own actions decided the outcome.

The views of Schopenhauer on the Will in nature work as well as theology (note that 'scientific laws' blend into 'will' in that system, and that cosmic bodies, or even e.g. the biosphere, in Bennett's system have 'will', but are not alive). Our model keeps the data intact from false design arguments, and also from reductionist flat history causal pseudo-arguments. We need hardly remind ourselves that the Old Testament has little value as an historical document, and was written very late. The Greek Archaic, as an analog, is a far better portrait of what's going on.

The modern era has reacted against the principles of spiritual powers because they are dangerous to freedom. The era of Feuerbach expressed this strongly. But even Kant noted in passing, after demolishing the design argument, the speculative hypothesis, after Plato, of a demiurge acting in history, the precursor of J. G. Bennett's 'demiurgic powers', an idea corrupted with occult additives, a warning of the danger of design arguments. We must remember that the Axial Age attempted (without success with the Israelites) to phase out polytheism and animal sacrifice. We must be vigilant to not let these phantoms return to haunt modernity. Bennett was a front man for rogue sufis trying to reinvent sacrifice and find a way to introduce a spiritual Trojan horse into modern ideology. Bennett came close to providing that. But his conception of world history, and of the rise of modernity are completely inadequate.

We must be wary of attempts to create religions of the Axial type all over again in the modern downfield.

We must remember that Christianity, Islam, Judaism, and Mahayana appear outside the Axial Age, and require another explanation, although the seeding analogy can work. But our distinction of system action and free action applies. These religions do not show 'process' but are generated in mideonic 'free agency'. But free agency can be spiritual. In any case, how do we explain the onset of Christianity, a very obscure data chaos?

The idea of demiurgic powers as Bennett's botched account shows is braided with a Trojan horse of sacrificial powers above man, the archaic nonsense all over again. Shadowy new age figures would enjoy undoing this achievement, and recreating the sacrificial. We should speak of the Kantian demiurge, singular or plural, of the Heavenly Host. These are speculative notions, from the first. Since we have no knowledge here, we might remain silent, save to put the 'god' idea in its supernatural context, leaving the issues of 'nature' to the lesser powers. Note that this merry-go-round has promptly reinvented near polytheism, and the way Christians displaced this into the realm of angelic powers.

> The slot for religious emergentism in the wake of the Axial interval passes in the modern transition into the aggressive revolutionary track of the modern atheist far left. Perhaps this was too extreme, but the point is clear, a new era has arrived to displace the Axial Age. But modern atheism has failed to grasp the option of material powers higher than man, a theme reborn in science fiction.

There is a plan B alternate, the Hegelian cult of the FrankenGeist, with a sideshow of the Owl of Minerva. In any case, the 'triad of Kant, Hegel, Schopenhauer' is a mysteriously apt end and new beginning to the Reformation. But the Feuerbachian way of slamming the religious door shut was perhaps inevitable at the dawn of a new era. Historical materialists emerged with a new 'superstition', the material dialectic, their 'ark of the covenant' conveying the ghost of '*Samkhya*' into a new future.

Understanding the Axial Age requires first giving a history of sacrifice as a religious practice, a history we can't really provide, to see its attempt to transcend beyond this bloody legacy, and its surreptitious alternates. We should note the tension in the Israelite transition as it failed to escape the gravitation of its basically Canaanite sources (and its Sumerian/Abrahamic mystery of ur-monotheism) with its legacies of animal sacrifice. The transition beyond sacrifice was thus accomplished by the later phases of Christianity (where the symbolism is superceded in the Christological card trick) and Islam (and Buddhism), and those two religions do not actually occur in the Axial Age. We are left with a dark question, what were all these fancies

Appendix

spiritual powers doing in eras before the Axial Age in dealing with a bloody-thirst humanoid, sapient at his best, at worst a cannibal.

Secular ideology has trended to exclude religion from modernist foundationalism, but this forgets the Reformations of the early modern, and this also forgets the complex issues of 'soul' that arise in early Christianity, and then again in Islam, which either spawns or hosts a complex formation called 'Sufism' whose exact parameters (and relation to soul questions) are obscure indeed.

> Our discussion moves beyond religion to the exotic and beautiful realm of the Kantian thematics of the idea of freedom, and shows the exotic case of 'freedom's causality'.

The world of Buddhism should be distinguished from Mahayana whose connection to the 'savior' religion of Christianity remains unresolved. The modern Reformation, which occurs early only to be sublated into a secular format, should remind us to be wary of the 'Israelite Reformation': the basic format of a Middle Eastern religious corpus undergoes a Reformation and then becomes something quite different. And this is backdated to the original forms, which we can see were the object of many attacks by the Prophets. The complexity of Israelitism is considerable indeed.

Our difficulties are all too reminiscent of the metaphysical issues raised in Kant, and we have seen a remarkable suggestion of the distinction of phenomenal directionality as representation of noumenal teleology. The idea of action from the virtual future (and in J. G. Bennett there is the idea of a hyparchic future) is related to that of teleology which has no place in standard science. We have in any case produced a solution to 'Kant's Challenge', and an insight into the relationship of history to evolution. This success of our model works even given the ragged character of its application, and its indifference to the incomplete character of the data. Flat history fails, what is the simplest alternative? The robust insight of a model as entry level as a finite transition series is remarkable to say the least: we are on the right track. In as important to critique such models as it is to attempt their use.

The model of flat history now dominant has completely misled secular thought. We see the clear operation of a higher power in history, but this does not validate theistic historicism and claims for divine revelation. We have replaced this with a systems analysis, but we have also explored the hypothesis of higher powers in nature, from biospheric or *Gaian* speculations to the Platonic/Kantian idea of the demiurge. The legacy of reductionist scientism is not able to solve the problem of the Axial Age, next to the equal confusions of theology. The simplest and best replacement for higher powers is the idea of teleological action, which seems to act from the future, but this still beggars the clear 'design equivalent' as 'mind in nature' that 'seems' to act throughout the Axial effects spectrum. This factor 'scans' culture streams and reamps certain factors as virtual categories or genres for human realization. The idea of teleology raises as many questions as it solves and we consider what is missing in the frameworks of science, and physics: as with our citation of a 'Goldilocks principle', the issue of ends invokes that of beginnings.

The Israelites deserve a little credit, even as we go on our secular path. They exposed factors in history that are hard to explain in normal causal terms. They were right to do so, and they noticed this happenstance as design as it unfolded in their Axial transition. The Greeks couldn't manage (but they thus did not confuse their history with mythologies of 'revelation'). The Israelites could not quite see how our 'stream and sequence' argument could have helped them sort our a revealing contradiction: the difference between the lead-up history (mostly myth) and the Axial interval history. They simply ignore that monotheism appears with Abraham but is created during the Axial period. There two kingdoms succumb to conquest, and a period of Exile blends two monotheisms, semitic and indo-european. These conquests precipitate globalization. The whole sequence of the Axial differential is spooky to say the least. But the whole Axial shows this kind of strangeness.

Consider this in light of the Greek Axial history of literature from Homer to Greek tragedy. Our depiction is literally true in the way a new genre of epic and tragedy generate human realizations. Something can scan the stream and amplify latent factors. And inside the Axial interval or its immediate post-divide point. Move from there to the literary 'epic' of the Israelites, the Old Testament. We've replaced religious superstition with science, no? We are stuck trying to reconstruct a 'design' action Frankenstein-style. If it can scan whole culture streams shouldn't we reinvent the category of pagan idols? If it can act like feedback over tens of millennia, shouldn't we allow a bit of theistic awe? The superstitious Israelites could still get the last laugh here, the joke on scientism. The buddhist case is even more spooky. Frankenstein generates an atheist religion. Like the mad hatter we just move on to the Chinese case. We can become Taoist nature mystics, and explore the topology of the Great Wall of China. Scientists are back to the wall here, but the situation is not hopeless. There are too many hints of simple

Appendix

dynamics here to indulge in mystic retrogression. But if, pace Frankenstein (and Hegel) the monstrous creation of theory becomes 'god', what then?

Our stance might simply adopt the method of the 'design inference' of the religion critics of Darwin: the Axial Age of Israel/Persia (and the rest) allows us to infer design, but this contradicts the properties of theistic 'god' due to its failure to exhibit omnipotence: it is a very limited natural system. We thus infer a power within nature that is behind the 'Axial Age', or ages. This puts finger in a very dangerous dyke: superstition recreation of religion-based on the extended 'Axial' or macro effect. Since the model shown hits a home run on its first time at bat, we suspect that such august higher powers are very good with hyper-advanced dynamical systems, macroevolution in our sense being one of them.

The irony here is that seeing the rise of modernity as a 'subsequent' Axial Age gives an example at close range of the type of transformation seen in fuzzy antiquity, and also explicates the riddle of modernity itself. The framework shows a sudden transition of three centuries capped with a 'divide' and the evidence fits beautifully here. The period after 1800, plus or minus, works perfectly as the onset of a new era in world history, after an explosive transition. The sudden transformation of economic systematics matched with the Industrial Revolution creates an entire new form of civilization.

But the issue in our model is not 'civilization' as such, the latter being a loose descriptive term, but our new 'unit' of analysis, the differential time-slice, a transition, whose effect generates a field of diffusion, or a new oikoumene. This time that oikoumene is truly global for the first time, and drama of economic globalization in many ways replaces the religious version of antiquity. But this crystallization as capitalism is really, to a close look, matched with the onset of a challenger in the rise of socialism and communism. The real challenger is the democratic revolution which operates an antithesis in the dialectic of freedom, and the counterpoint of free markets and the socialist freedom of individuals or proletariats shows the real dynamic. In a period two centuries downfield the significance of this larger 'dialectic' is becoming apparent and the final stage of the world economy will return to its cultural integration via a secular version of religion.

But we should note that modernity generates a Reformation, and this is as secular as anything else, and is a prime generator of the democratic revolution and of abolition. If we examine the 'Israelite' transition in antiquity we see one and the same process: a 'reformation' of middle eastern or Canaanite religion, followed by a lot of backdated history tending to conceal this onset, in a process similar to the modern. But the modern transition shows also a continuation from the Axial Age religions in reformation, and the final outcome here is still in the making. We need to point out that this is not the same as the triumphant domination of 'science', but the balanced spectrum dialectic of modernity with its highly complex dynamic of the causal and the idea of freedom.

> It is important to see how the Enlightenment produces a morality for the first time in Kant's remarkable *Critique of Practical Reason*. However open to dialectical challenge it is a remarkable first. Moses and his Tablets was a myth of primitive religion...

The modern world is reacting against the theism of the ancient Axial Age, despite the continuity of the Reformation. The prompt appearance of atheism as a world view begins in the seventeenth century. But the confusions of this with scientism and the entanglement with reductionist evolutionism has given the field of religious conservatives a stay. The issue of evolution was better resolved in the early modern with the work of figures like Kant and his generation of the school of teleomechanists, a reminder of the frequent onset of confusion in the wake of our mysterious transitions. The problem shows a sign of a solution in the reformulation of 'evolution' in our model.

The modern transition is really the conclusion to the classic Axial Age, and the epochs before it, and it seems unlikely that the macrosequence will continue to another in the future: man appears to have graduated from the second phase of The Great Transition and as 'system action' yields to 'free action' his confronted for a task that is so far beyond his powers, the problem of understanding let alone replicating an evolutionary transformation on the scale of tens of millennia. By comparison, the only analog, the modern revolution, is downright primitive. And yet the analog suggests something about the future prospects of humanity. Man cannot simply drift with the tides of successive epochs ending up rescued from medieval worlds by the coming of a future transformation. He is confronted with a task of great complexity, and so far, beyond his powers, yet one that will subject him to the test of a real future.

We are confronted with an emergence into freedom with an existential suspense involved in being the inheritor of the gifts of nature almost before we are ready to receive them. We enter the treacherous realm of man's evolution becoming the self-evolution of man, armed with a darwinian view of man that is inadequate to the reality we now see. But man is entering slowly the era of technological terraforming, and the companion discoveries must be of the nature of social change itself, and the prospect of generating social processes on the scale of civilization itself. These are not the same as elements of capitalist economy that have become a probably temporary phase of modernity. Current social consciousness thinks nothing of eliminating the Amazon basin in the sprawl of capitalist dynamics. Clearly, the first phase of modernity is confusing us.

We need to see that as *homo sapiens* emerges from the Great Transition, or, equally, his entry into its final phase, his responsibility for a whole planet demands a cogent self-evolution into creative self-consciousness, in the realization of his real nature in a Gaian context, and this new man, the last of the first men, or the first after the last men, must recover the natural state bestowed on man at the dawn of man, a state analogous to the buddhist or yogi's 'enlightenment', in the realization of who he was from the start. This is the real succession to the era of world religions now drawing to a close, but with a cautionary cross-examination of the secular era coming to the fore. Man has received a great deal of help in the creation of civilization, his passage to his real autonomy is an adventure demanding the transformation of man as he is, and a that will be a recovery of the first men at the dawn of *homo sapiens*.

As we look back from the stance of the 'modern' world we see that the hypothesis of the modern transition works beautifully as an account of the rise of a new era of world history, and we are still in the early phases of that successor epoch. We see that the Reformation transformed Christianity, while the legacy of Buddhism crept into modernity almost as early as the Enlightenment itself, generating a comic *double entendre* of terms, one that will transform modernity into multiple melodies. But the revolutionary left usurped the place of religion in the period just following the modern 'divide' and the rich man's religion of the Calvinists, next to the suppression of the primordial communism of the Christian rebel Thomas Münzer, will pass away before the efforts to bring the modern proletariats to full inclusion in basic civilization. The resemblance to the underground revolt of the Christians is no accident, but the reversal into materialist ideology will force the issue of metaphysical 'reboot'. The materialism of the early marxists will no doubt yield to something kin to the implications of 'spiritual *Samkhya* materialism' of the ancient yogis, or else the efforts of neuroscience to reckon with simple consciousness, the ultimate endgame science of the progeny of Newtonian fundamentalism.

The way forward was clearly foreseen by the philosopher Kant but the science as myth of the early positivists of scientism will prove a long dialectical struggle of science with itself. Modernity gave birth to the idea of freedom (in reality another Greek idea) and its implications in relation to science confound the regime of mechanism with first the biological, and then the cosmological. Christianity was in many ways a political revolution of the Roman Empire and the regime of slavery, the motion of the modern democratic revolution to the conclusions of socialism and communism will likewise answer to the globalization of capitalism. Behind the disguises of time and culture the resemblance to the antique Axial instances is remarkable. The Great Transition seems to be yielding to the maturity of man becoming man, and the permutation of the catch phrase 'last and first men', suggests that first real civilization will be as a construct of the first men against the ages of the last.

5. BIBLIOGRAPHY

Adams, Barbara. *Predynastic Egypt*. UK: Shire, 1988.
Aldred, Cyril. *Egypt to the End of the Old Kingdom*, New York: McGraw Hill, 1965.
Allen, Robert. *Global Economic History: A Short Introduction*. New York: Oxford University Press, 2011.
Allen, Robert. *The British Industrial Revolution in Global Perspective*. New York: Cambridge University Press, 2009.
Andress, David. *The Terror: The Merciless War For Freedom in Revolutionary France*. New York: Farrar, Straus, And Giroux, 2005.
Anthony, David. *The Horse, the Wheel and Language: How Bronze-age Riders from the Eurasian Steppes Shaped the Modern World*. Princeton: Princeton University Press, 2007.
Arblaster, Anthony. *The Rise and Decline of Western Liberalism*. New York: Basil Blackwell, 1984.
Armstrong, Karen. *The Great Transformation: The Beginning of Our Religious Traditions*. New York: Knopf, 2006.
Balibar, Etienne. *The Philosophy of Marx*. New York: Verso, 2007.
Balter, Michael. *The Goddess and the Bull: Çatalhöyük, an Archaeological Journey to the Dawn of Civilization*. New York, Free Press, 2005.
Bannister, Robert. *Social Darwinism: Science and Myth in Anglo-American Thought*.

Philadelphia: Temple University Press, 1979.
Baring, Ann, et al. *The Myth of the Goddess: Evolution of an Image*. New York: Viking Arkana, 1991.
Barnes, Jonathan. *Early Greek Philosophy*. New York: Penguin, 2001.
Barrow, John et al. *The Anthropic Cosmological Principle*. New York: Oxford, 1986.
Barzun, Jacques. *Darwin, Marx, Wagner: Critique of a Heritage*. Boston: Little, Brown, 1941.
Barzun, Jacques. *From Dawn to Decadence: 500 Years of Western Cultural Life 1500 to the Present*. New York: HarperCollins, 2000,
Basham, A. L. *The Wonder that was India: A Study of the History and Culture of the Indian Sub-continent Before the Coming of the Muslims*. New York: Taplinger, 1967.
Bauer, Susan Wise. *The History of the Ancient World: From the Earliest Accounts to the Fall of Rome*. New York: Norton, 2007.
Beckwith, Christopher. *Empires of the Silk Road: A History of Central Eurasia from the Bronze Age to the Present*. Princeton: Princeton University Press, 1009.
Behe, Robert. *Darwin's Black Box: The Biochemical Challenge to Evolution..* New York: Free Press, 1996.
Bender, Barabara. *Farming in Prehistory: From Hunter-gatherer to Food-producer*. New York: St Martin's Press, 1975.
Billington, James. *Fire in the Minds of Men: Origins of the Revolutionary Faith*. New York: Basic Books, 1980.
Blanning, Tim. *The Pursuit of Glory: The Five Revolutions That Made Modern Europe 1648-1815*. New York: Penguin, 2008.
Boas, Marie. *The Scientific Renaissance: 1450-1630*. New York: Harper, 1962.
Boner, Harold. *Hungry Generations, The Nineteenth-Century Case Against Malthusianism*. New York: King's Crown Press, 1955.
Bowler, Peter. *Evolution: History of an Idea*. Berkeley: University of California Press, 2003.
Brackman, Arnold. *A Delicate Arrangement: The Strange Case of Charles Darwin and Alfred Russell Wallace. New York*: Times Books, 1980.
Brown, Cynthia Stokes. *Big History: From the Big Bang to the Present*. New York: The New Press, 2007.
Buck-Morss, Susan. *Hegel, Haiti, and Universal History*. Pittsburgh, Penn.: Pittsburgh University Press, 2009.
Buhle, Paul. *Marxism in the United States: A History of the American Left*. New York: Verso, 2013.
Burns, A. R. *The Lyric Age of Greece*. New York: St. Martin's, 1960.
Carroll, Sean et al. *From DNA to Diversity: Molecular Genetics and the Evolution of Animal Design*. New York: Blackwell, 2001.
Cavalli-Sforza, Luigi. *The Great Human Diasporas: The History of Diversity and Evolution*. New York: Addison-Wesley, 1993.
Ceram, C. W. *Gods, Graves, and Scholars*. New York: Knopf, 1967.
Chalmers, Johnson. *Nemesis: The Last Days of the American Republic*. New York: Henry Holt, 2006.
Chang, Kwang-chih. *The Archaeology of Ancient China*. New Haven: Yale University

Press, 1977.
Charles Freeman. *Egypt, Greece and Rome: Civilizations of the Ancient Mediterranean.* New York: Oxford, 1991.
Childe, Gordon. *Man Makes Himself.* New York: New American Library, 1951.
Clagett, Marshall. *Greek Science in Antiquity.* New York: MacMillan, 1955
Coe, Michael. *The Maya.* New York: Thames and Hudson, 1966.
Cockshott, Paul et al. Towards a New Socialism. Nottingham, UK: Spokesman, 1993.
Cohen, H. Floris. *The Scientific Revolution: A Historiographical Enquiry.* Chicago: University of Chicago Press, 1994.
Cohen, I. Bernard. *Revolution in Science.* Cambridge: Harvard University Press, 1985.
Cohn, Norman. *Cosmos, Chaos and the World to Come*: *The Ancient Roots of Apocalyptic Faith.* New Haven: Yale University Press, 1993.
Comay, Rebecca. *Mourning Sickness: Hegel and the French Revolution.* Standford University Press. Stanford, Ca., 2011.
Coote, Robert. *Early Israel: A New Horizon.* Minneapolis: Fortress, 1990.
Corsi, Pietro. *The Age of Lamarck: Evolutionary Theories in France, 1790-1830.* Berkeley: University of California Press, 1988.
Crawford, Harriet. *Sumer and the Sumerians.* New York: Cambridge University Press, 1991.
Creel, H. G. *The Birth of China:* A *Study of the Formative Period of Chinese Civilization.* New York: Frederick Ungar, 1967.
Curtis, Gregory. *The Cave Painters: Probing the Mysteries of the World's First Artists.* New York: Anchor, 2006.
Danielou, Alain. *A Brief History Of India*, Rochester, Vermont: 2003.
David Damrosch. *The Buried Book: The Loss and Rediscovery of the Great Epic of Gilgamesh.* New York: Henry Holt, 2006.
Davies, Norman. *Europe: A History.* New York: HarperPerennial, 1998.
Davis, Brion Davis. *The Problem of Slavery in Western Culture.* New York: Oxford University Press, 1966.
Dawkins, Richard. *Climbing Mount Improbable.* New York: Norton, 1996.
Dawkins, Richard. *The Blind Watchmaker.* New York: Norton, 1987.
Dawkins, Richard. *The God Delusion.* New York: Houghton-Mifflin, 2006.
de Chardin, Teilhard. *The Phenomenon of Man.* New York: Harper and Row, 1959.
De Ste. Croix, G.E.M. *The Class Struggle in the Ancient Greek World.* Ithaca, NY: Cornell University Press, 1981.
de Vaux, Roland. *The Bible and the Ancient Near East.* New York: Doubleday, 1971.
Dembski, William. *No Free Lunch.* New York: Rowman & Littlefield, 2002.
Dennett, Daniel, *Darwin's Dangerous Idea: Evolution and the Meanings of Life.* New York: Simon & Schuster, 1995.
Dennett, Daniel. *Freedom Evolves.* New York: Viking, 2003.
Denton, Michael. *Evolution: A Theory in Crisis.* New York: Adler & Adler, 1985.
Desmond, A. et al. *Darwin, Life of a Tormented Evolutionist.* New York: Warner, 1991.
Derman, Emanuel. *Models, Behaving Badly: Why Confusing Illusion with Reality Can Lead to Disaster, on Wall Street and in Life.* New York: Free Press, 2011.
Dever, William. *Who Were The Israelites and Where Did They Come From?* Grand

Rapids, Michigan: Eerdmans, 2003.
Dickerson, Derek. *Adam's Tongue: How Humans Made Language, How Language Made Humans.* New York: Hill & Wang, 2009
Dodds, E. R. *The Greeks and the Irrational.* Berkeley: University of California Press, 1951.
Donald Redford. *Egypt, Canaan, and Israel in Ancient Times.* Princeton: Princeton University Press, 1992.
Doyle, William. *Origins of the French Revolution.* New York: Oxford University Press, 1980.
Eiseley, Loren. *Darwin's Century: Evolution and the Men Who Discovered* It. New York: Anchor Books, 1961.
Eldredge, Niles. *The Myths of Human Evolution.* New York: Columbia University Press, 1982.
Else, Gerald. *The Origin and Early Form of Greek Tragedy.* Cambridge: Harvard University Press, 1967.
Emery, Walter. *Archaic Egypt.* New York: Penguin, 1961.
Erlichman, Howard. *Conquest, Tribute, and Trade: The Quest For Precious Metals and the Birth of Globalization.* Amherst, NY: Prometheus, 2010
Farrington, Benjamin. *Greek Science: Its Meaning for Us.* New York: Penguin, 1949.
Faulkner, Neil. *A Marxist History of the World From Neanderthals to Neoliberals.* London: Pluto Press, 2013.
Ferrill, Arther. *The Fall of the Roman Empire.* New York: Thames and Hudson, 1986.
Finkelstein, I. et al. *The Bible Unearthed.* New York: The Free Press, 2001.
Finley, Moses. *Ancient Slavery and Modern Ideology.* New York: Penguin, 1983.
Finley, Moses. *Early Greece: The Bronze and Archaic Ages.* New York: Norton, 1981.
Foley, Duncan. *Adam's Fallacy: A Guide to Economic Theology.* Cambridge: Harvard University Press, 2006.
Forrest, W. G. *The Emergence of Greek Democracy: 800 to 400 BC.* New York: McGraw-Hill, 1966.
Foster, Benjamin et al. *Civilizations of Ancient Iraq.* Princeton: Princeton University Press, 2009.
Fox, Robert Lane. *The Classical World: An Epic History From Homer to Hadrian.* New York: Basic Books, 2006.
Fox, Robert Lane. *The Unauthorized Version: Truth and Fiction in the Bible.* New York: Knopf, 1992.
Freeman, Charles. *The Closing of the Western Mind: The Rise of Faith and the Fall of Reason.* New York: Random House, 2002.
Friedman, Richard. *Who Wrote the Bible?* New York: Summit Books, 1987.
Fukuyama, Francis. *The End of History and the Last Man.* New York: The Free Press, 1992.
Furet, Francois. *Interpreting the French Revolution.* Cambridge: Cambridge University Press, 1978.
Garbini, Giovanni. *History and Ideology in Ancient Israel.* London: SCM, 1988.
Gardiner, Sir Alan. *Egypt of the Pharaohs: An Introduction.* New York: Oxford University Press, 1964.

Bibliography

Gay, Peter. *The Enlightenment: An Interpretation*. New York: Norton, 1966.
Gerhart, J. et al. *Cells, Embryos, and Evolution: Toward a Cellular and Developmental Understanding of Phenotypic Variation and Evolutionary Adaptability*. New York: Blackwell, 1997).
Gernet, Louis. *The Anthropology of Ancient Greece*. Baltimore: Johns Hopkins University Press, 1981.
Gillespie, Neal. *Charles Darwin and the Problem of Creation*. Chicago: The University of Chicago Press, 1979.
Goldsworthy, Adrian. *How Rome Fell: Death of a Superpower*. Newe Haven: Yale University Press, 2009.
Gordon, Cyrus. *Before Columbus; Links Between the Old World and Ancient America*. New York: Crown, 1971.
Gould, S. J. *The Structure of Evolutionary Theory*. Cambridge: Harvard University Press, 2002.
Gould, S. J. *Time's Arrow, Time's Cycle*. Cambridge: Harvard University Press, 1987.
Greene, John. *Science, Ideology, and World View*. Berkeley: University of California Press, 1981.
Hall, A. Rupert. *The Revolution in Science 1500-1750*. New York: Longman, 1983.
Hampson, Norman. *A Cultural History of the Enlightenment*. New York: Pantheon, 1968.
Harris, Leon. *Evolution: Genesis and Revelations, With Readings from Empedocles to Wilson*. Albany: State University of New York Press, 1981.
Hartl, D. et al. *Principles of Population Genetics*. Sunderland, Mass: Sinauer Associates, 1997.
Harvey, David. *A Brief History of Neoliberalism*. New York: Oxford University Press, 2005.
Harvey, David. *A Companion to Marx's Capital*. New York: Verso, 2010.
Hayek, F. A. *The Counterrevolution of Science*. London: Collier-MacMillan, 1955.
Hazard, Paul. *The European Mind: 1680-1715*. New York: Fordham, 1990.
Heather, Peter. *The Fall of the Roman Empire: A New History Of Rome and the Barbarians*. New York: Oxford University Press, 2006.
Henry, Donald. *From Foraging to Agriculture: The Levant at the End of the Ice Age*. Philadelphia: University of Pennsylvania, 1989.
Hill, Christopher. *The English Bible and the Seventeenth-Century Revolution*. New York: Penguin, 1993.
Hill, Christopher. *The World Turned Upside Down: Radical Ideas During the English Revolution*. New York: Penguin, 1991.
Hobsbawm, Eric. *The Age of Revolution: 1789-1848*. New York: New American Library, 1962.
Hodgson, Marshall. *Rethinking World History: Essays on Europe, Islam and World History*. Cambridge: Cambridge University Press, 1993.
Hofstadter, Richard. *Social Darwinism in American Thought*. Philadelphia: University of Pennsylvania Press, 1945.
Hoyle, F. et al. *Evolution From Space*: *A Theory of Cosmic Creationism*. London: Dent, 1981.

Hughes, Stuart. *Oswald Spengler: A Critical Estimate*. New York: Charles Scribner, 1952.
Hunt, Lynn. *Inventing Human Rights: A History*. New York: Norton, 2007.
Huxley, Julian. *Evolution,The Modern Synthesis*. New York: Allen & Unwin, 1942.
Huxley, T. H. *Evolution and Ethics*. Princeton: Princeton University Press, 1989.
Israel, Jonathan. *Radical Enlightenment: Philosophy and the Making of Modernity 1650-1750*. Oxford: Oxford University Press, 2002.
Jacquette, Dale. *The Philosophy of Schopenhaur*. McGill-Queen's University Press, 2005.
Jaspers, Karl. *The Origin and Goal of History*. New Haven: Yale University Press 1953
Johnson, Philip. *Darwin on Trial*. Downers Grove, Ill.: InterVarsity, 1993.
Jones, E. L. *Growth Recurring: Economic Change in World History*. Clarendon:Oxford University Press, 1988.
Jones, E. L. *The European Miracle: Environments, Economies, and Geopolitics in the History of Europe and Asia*. Cambridge: Cambridge University Press, 1981.
Kagan, Donald. *Pericles of Athens and the Birth of Democracy*. New York: The Free Press, 1991.
Kauffman, Stuart. *At Home in the Universe: The Search for Laws of Self-organization and Complexity*. New York: Oxford University Press, 1995.
Kelly, Kevin. *Out of Control: The Rise of Neo-biological Civilization*. New York: Addison-Wesley, 1994.
King-Hele, Desmond. *Erasmus Darwin: Grandfather of Charles Darwin*. New York: Scribners, 1963.
Kirschner, M. et al. *The Plausibility of Life: Resolving Darwin's Dilemma*. New Haven: Yale University Press, 2005.
Klein, Richard et al. *The Dawn of Human Culture*. New York: Wiley, 2002.
Koestler, Arthur. *The Sleepwalkers; A History of Man's Changing Vision of the Universe*. New York: Grosset & Dunlap, 1959.
Körner, Stephan. *Kant*. New York: Penguin, 1962.
Kosambi, Damoda. *Ancient India; A History of Its Culture and Civilization*. New York: Random House, 1965.
Kramer, Samuel. *The Sumerians*. Chicago: Chicago University Press, 1963.
Landes, David. *The Unbound Prometheus: Technological Change and Industrial Development in Western Europe from 1750 to the Present*. New York: Cambridge University Press, 1969.
Lane, Nick. *Life Ascending: The Ten Great Inventions of Evolution*. New York: Norton, 2009.
Larson, Edward. *Evolution: The Remarkable History of a Scientific Theory*. New York: The Modern Library, 2004.
Lebowitz, Michael. *Build It Now: Socialism for the Twenty-First Century*: New York: Monthly Review Press, 2006.
Leech, Garry. *Capitalism: A Structural Genocide*. London: Zed Books, 2012.
Lemche, N. P. *Ancient Israel, A New History of Israelite Society*. Sheffield, England:

JSOT Press, 1988.
Lenoir, Timothy. *The Strategy of Life: Teleology and Mechanics in Nineteenth-Century German Biology*. Chicago: University of Chicago Press, 1989.
Lovelock, James. *Gaia: A New Look at Life on Earth*. New York: Oxford University Press, 1979.
Lovtrup, Soren. *Darwinism: Refutation of a Myth*. New York: Croom Helm, 1987.
Lyons, Sherrie. *Thomas Henry Huxley: The Evolution of a Scientist*. New York: Prometheus, 1999.
McKibben, Bill. *The End of Nature*. New York: Random House, 1989.
MacNally, David, (et al.), *Catastrophism: The Apocalyptic Politics of Collapse and Rebirth*, Oakland, Ca.: PM Press, 2012.
MacNally, David. *Monsters of the Market: Zombies, Vampires, and Global Capitalism*. Chicago, Il.: Haymarket, 2011.
MacNeill, William. *The Rise of the West*. Chicago: University of Chicago Press, 1963.
Magee, Bryan. *The Philosophy of Schopenhauer*. New York: Clarendon, 1997.
Mah, Harold. *The End of Philosophy, the Origin of 'Ideology'*. Berkeley: University of California Press, 1987.
Maisels, Charles. *The Emergence of Civilization: From Hunting and Gathering to Agriculture, Cities, and the State in the Near East*. New York: Routledge, 1990.
Majumdar, R. C. *The History and Culture of the Indian People*. London: George Allen, 1951.
Mallory, J. P. *In Search of the Indo-Europeans: Language, Archaeology, and Myth*. New York: Thames & Hudson, 1989.
Mazlish, Bruce. *The Riddle of History; The Great Speculators from Vico to Freud*. New York: Harper & Row, 1966.
McFarland, J. D. *Kant's Concept of Teleology*. Edinburgh: University of Edinburgh Press, 1970.
McLaughlin, Peter. *Kant's Critique of Teleology in Biological Explanation*. Lewisten, New York: Edwin Mellen, 1990.
McNeill, William. *The Rise of the West*. Chicago: Chicago University Press, 1963.
Mellaart, James. *Earliest Civilizations of the Near East*. London: Thames & Hudson, 1965.
Mertz, Barbara. *Temples, Tombs, and Hieroglyphs: A Popular History of Ancient Egypt*. New York: Dodd, Mead, 1978.
Miller, G. et al. *Origination of Organismic Form*. Cambridge: MIT Press, 2002.
Miller, James. *Rousseau, Dreamer of Democracy*. New Haven: Yale University Press, 1984.
Mithen, Steven. *After The Ice: A Global Human History, 20,000-5000 BC*. Cambridge: Harvard, 2004.
Murray, Oswyn. *Early Greece*. Cambridge: Harvard University Press, 1993.
Needham, Joseph. *Science and Civilization in China*. Cambridge: Cambridge University Press, 1965.
Nicholas Wade. *The Faith Instinct: How Religion Evolved and Why it Endures*.

New York: Penguin. 2009.
Nissen, Hans. *The Early History of the Ancient Near East*. Chicago: University of Chicago Press, 1988.
O'Donnell, James J. *The Ruin of the Roman Empire: A New History*. New York: HarperCollins, 2008.
Ober, Josaiah. *Demokratia: A Conversation on Democracies, Ancient and Modern*. Princeton: Princeton University Press, 1996.
Olmstead, A.T. *History of the Persian Empire*. Chicago: Chicago University Press, 1948.
Olson, Richard. *Science and Scientism in Nineteenth Century Europe*. Chicago: University of Illinois Press, 2008.
Olson, Steve. *Mapping Human History: Genes, Race and Our Common Culture*. New York: Mariner Books. 2003.
Oppenheimer, Stephen. *The Real Eve: Modern Man's Journey Out of Africa*. New York: Carrol & Graf, 2003.
Panitch, Leo (et al.), *The Making of Global Capitalism: The Political Economy of American Empire*, New York: Verso, 2012.,
Paradis, James. *T. H. Huxley: Man's Place in Nature*. Lincoln: University of Nebraska Press, 1978.
Parenti, Christian. *Tropic of Chaos: Climate Change and the New Geography of Violence*. New York: Nation Books, 2011.
Peel J. D. *Herbert Spencer: The Evolution of a Sociologist*. New York: Basic Books, 1971.
Pennock, Robert (ed.) *Intelligent Design Creationism and Its Critics: Philosophical, Theological, and Scientific Perspectives*. Cambridge, Mass.: MIT Press, 2001.
Pennock, Robert. *The Tower Of Babel: The Evidence against the New Creationism*. Cambridge: The MIT press, 1999.
Pohlenz, Max. *Freedom in Greek Life and Thought*. New York: Humanities Press, 1966.
Polanyi, Karl. *The Great Transformation*. Boston: Beacon Hill, 1957.
Popper, Karl. *The Poverty of Historicism*. New York: Routledge, 1991.
Provine, William. *The Origins of Theoretical Population Genetics*. Chicago: University of Chicago Press. 1971.
Raff, Rudolf. *The Shape of Life: Genes, Development, and the Evolution of Animal Form*. Chicago: University of Chicago, 1996.
Redford, Donald. *Egypt, Canaan, and Israel in Ancient Times*. Princeton: Princeton University Press, 1992.
Redman, Charles, *The Rise of Civilization: from Early Farmers to Urban Society in the Ancient Near East*, San Francisco: Freeman, 1978
Reid, Robert. *Evolutionary Theory, The Unfinished Synthesis*. New York: Cornell University Press, 1985.
Rice, Michael. *Egypt's Making: The Origins of Ancient Egypt*. New York: Routledge, 1990.
Roberts, J. M., *The Penguin History of the World*, New York: Penguin, 1995

Roux, George. *Ancient Iraq*. New York: Penguin, 1992.
Ruse, Michael. *The Darwinian Revolution: Science Red in Tooth and Claw*. Chicago: University of Chicago Press, 1999.
Ryan, Frank. *Darwin's Blind Spot: Evolution Beyond Natural Selection*. New York; Houghton Mifflin. 2002.
Saggs, H.W.F. *Civilization Before Greece and Rome*. New Haven: Yale University Press, 1989.
Schiavone, Aldo. *The End of the Past: Ancient Rome and the Modern West*. Cambridge: Harvard University Press, 2000.
Schwarz, Jeffrey. *Sudden Origins: Fossils, Genes, and the Emergence of Species*. New York: Wiley, 1999.
Secord, James. *Victorian Sensation: The Extraordinary Publication, Reception, and Secret Authorship of Vestiges of the Natural History of Creation*. Chicago: University of Chicago Press, 2003.
Sklar, Judith. *Men and Citizens: A Study of Rousseau's Social Theory*. New York: Cambridge University Press, 1969.
Starr, Chester, *The Economic and Social Growth of Early Greece: 800-500 B.C.*, Oxford: Oxford University Press, 1977
Starr, Chester. *The Origins of Greek Civilization*. New York: Norton, 1991.
Sykes, Brian. *The Seven Daughters of Eve: The Science That Reveals Our Genetic Heritage*. New York: Norton, 2001.
Taha, Abir. *Nietzsche: Prophet Of Nazism: The Cult of the Superman, Unveiling the Nazi Secret Doctrine*. Bloomington, Indiana: Author House, 2005.
Tallis, Raymond. *Aping Mankind: Neuromania, Darwinitis, and the Misrepresentation of Humanity*. Durham, UK: Acumen, 2011).
Tucker, Robert. *Philosophy and Myth in Karl Marx*. New York: Cambridge University Press, 1964.
Van Seters, John. *In Search of History: Historiography in the Ancient World and the Origins of Biblical History*. New Haven: Yale University Press, 1983.
Vernant, Jean-Pierre. *Myth and Society in Ancient Greece*. New York: Zone Books, 1988.
Vogt, Joseph. *Ancient Slavery and the Ideal of Man*. Oxford: Basil Blackwell, 1974.
Wallerstein, Immanuel. *Historical Capitalism*. New York: Verson, 2011.
Ward, Keith. *The Development of Kant's View of Ethics*. New York: Blackwell, 1972.
Waterfield, Robin. *Why Socrates Died: Dispelling the Myths*. New York: Norton, 2009.
Watkins, Calvert. *How to Kill a Dragon: Aspects of Indo-European Poetics*. New York: Oxford University Press, 1995.
Weikart, Richard. *From Darwin to Hitler: Evolutionary Ethics, Eugenics and Racism in Germany*. New York: MacMillan, 2004.
Wells, Spencer. *The Journey Of Man: A Genetic Odyssey*. New York: Random House, 2003.
Wesson, Robert. *Beyond Natural Selection*. Cambridge: MIT, 1991.
Williams, Chris. *Ecology and Socialism: Solutions to Capitalist Ecological Crisis*. Chicago: Haymarket, 2010.

Witham, Larry. *Where Darwin Meets the Bible: Creationists and Evolutionists in America*. Oxford: Oxford University Press, 2002.

Woodhouse, W. J. *Solon The Liberator: A Study of the Agrarian Problem in Attika in the Seventh Century*. New York, Octagon, 1965..

Woolley, Leonard. *The Sumerians*. Oxford: Clarendon, 1928.

6. ILLUSTRATIONS

WC= Wikimedia Commons, PD= public domain CA = Common Attribution
Cover: WCPD: File:Leonardo Coccorante - 'Harbor Scene with Roman Ruins'.jpg
Back Cover WCPD File:'Cicadia and Vines' by Shibata Zeshin, Honolulu Museum of Art, 4653.1-3.jpg
TOC File:Storck A port in the south.jpg
Preface File:Leonardo Coccorante - 'Harbor Scene with Roman Ruins'.jp
5.WCPD File:Gutenberg press.jpg 6. WCPD Gérard - Painter when painting a portrait of a lute player.jpg 7. WCPD File:Cole Thomas The Course of Empire Desolation 1836.jpg

Preface
1. WCPD File:The Triumph of Civilization.jpg, 1793, Jacques Reattu
2. WCPD File:YHWH on Lakis Letters (no. 2).jpg
3. WCPD File:Karl Jaspers-BA.jpg
4. WCPD File:VajraMudra.JPG
5. WCPD File:Lao Tzu - Project Gutenberg eText 15250.jpg
6. WCPD File:Gustave Doré - The Holy Bible - Plate I, The Deluge.jpg
7. WCPD File:William Blake 008.jpg
8. WCPD File:Caspar David Friedrich 022.jpg
9. WCPD File:'Tirthankara', India, Mysore, Karnataka, 10th-11th century, bronze with silver content, Honolulu Academy of Arts.JPG
10. WCPD File:A young woman is sitting in a chair reading a story which ha Wellcome V0040287.jpg
11. WCPD File:Skrifmaskin, Smith Premier-maskin, Nordisk familjebok.png

12. WCPD File:Violin.JPG
13. WCPD File:Gebhard Fugel Moses erhält die Tafeln.jpg
14. WCPD File:Akenaton.jpg
15. WCPD File:August Wilhelm Julius Ahlborn - Blick in Griechenlands Blüte - Google Art Project.jpg
16. WCPD File:MedicineMan.Catlin.jpg
17. WCPD File:Athens ruins c1870 detail.jpg
18. WCPD File:Jacob-angel.jpg (Dore)
19. WCPD 11. File:Indischer Maler um 1660 002.jpg Krishna tanzt zur Musik zweier Mädchen
20. WCPD File:Lord Mahavir Gold.jpg
21. WCPD File:Tree.ring.arp.jpg
22. File:'Shepherd and Sheep' by Anton Mauve, Cincinnati Art Museum.JPG
23. WCPD WC CC 2.0 File:South Fork Citico Creek.jpg This file is licensed under the Creative Commons Attribution-Share Alike 2.0 Generic license. Chris M Morris/Wilderness/Flicker. Image cropped/BW rendered // This image was originally posted to Flickr by cm195902 at http://flickr.com/photos/79666107@N00/234084525. It was reviewed on 6 August 2010 by the FlickreviewR robot and was confirmed to be licensed under the terms of the cc-by-sa-2.0.
24. File:Cro-Magnon-female Skull.png
25. WCPD File:Richer - Premier Artiste 3.jpg
26. WCPD File:Dirc van Delft - God Bestowing a Soul on Adam - Walters W17125R - Full Page.jpg
27. WCPD File:William Powell as Hamlet encountering the Ghost (Wilson, c. 1768-1769).jpg
28. WCPD File:Shiva Pashupati.jpg
29. WCPD File:Egyptian - Ba Bird - Walters 571472 - Back.jpg
30. WCPD File:Maya vessel with sacrificial scene DMA 2005-26.jpg
31. WCPD File:Greekreligion-animalsacrifice-corinth-6C-BCE.jpg
32. WCPD File:M&U painting 3.jpg
33. WCPD File:Northrop Abraham Offers Isaac.jpg
34. WCPD File:Mayan priest smoking.jpg
35. WCPD File:Warriors2.jpg Mayan "Temple of the Warriors", Chicken Itza, Mexico
36. WCPD File:Zeus Yahweh.jpg

Introduction
Header: WCPD File:Eismann, Johann Anton - Küstenlandschaft mit antiken Ruinen.jpg
1.1. WCPD File:Zeus-Tempel von Eumonos.jpg
1.2. WCPD File:Heraclitus. Line engraving by B. S. Setlezky after G. B. Goe Wellcome V0002703.jpg
1.3. WCPD File:120.The Prophet Isaiah.jpg
1.4. WCPD File:Ascetic Bodhisatta Gotama with the Group of Five.jpg
1.5. WCPD File:Confucius Statue at the Yushima Seido.JPG
1.6. WCPD File:Cave painting, Anthropos (2).jpeg
1.7. WCPD File:Henry Mark Anthony. Stonehenge.jpg
1.8. WCPD File:'The Pyramids of Sakkarah from the North East'..jpg

Illustrations 289

1.9. WCPD File:Sumerian 26th c Adab.jpg
1.10. WCPD File:Jean-Baptiste Lamarck.jpg Lamarck
1.11. WCPD File:Charles Darwin by G. Richmond.jpg
1.12. WCPD File:Alfred Russel Wallace 1862 - Project Gutenberg eText 15997.png
1.13. WCPD File:PaestumItalien.jpg
1.14. WCPD File:Dodwell Parthenon 3.jpg
1.15. WCPD File:Hubert Sattler Kolosse des Memnon 1846.jpg
1.16. WCPD File:David Roberts - Philae.jpg
1.17. WCPD File:1911 Britannica-Arachnida-Sao hirsuta.png
1.18. WCPD File:EarlyPleistoceneAnimals.png
1.19. WCPD File:Anatomical and geometrical proportions - Albrecht Dürer.png
1.20. WCPD File:Kant-KdrV-1781.png

Chapter 2
Header WCPD File:Roemische Ruinenlandschaft Paul Bril.jpg
2.1 WCPD File:Official Photographs taken on the Front in France - View of Gommecourt as seen today (15560800766).jpg
2.2 WCPD File:The Second Battle of Ypres.jpg
2.3 WCPD File:Kant5.jpg
2.4 WCPD File:Jean-Francois Champollion 2.jpg
2.5 WCPDFile:Rosetta Stone.jpg
2.6 WCPD File:Cole Thomas The Course of Empire Destruction 1836.jpg
2.7 Mouse motif, drawn by author
2.8 WCPD File:Six-volume 'The Holy Land' by David Roberts.jpg
2.9 WCPD File:The divided kingdom.jpg
2.10 WCPD File:Cole Thomas Expulsion from the Garden of Eden 1828.jpg
2.11 WCPD File:Adam and Eve Driven out of Eden.png
2.12 WCPD File:Tissot The Flight of the Prisoners.jpg
2.1 WCPD File:Nazareth the holy land 1842.jpg roberts
2.14 WCPD File:Zoroaster 1.jpg
2.15 WCPD File:Patriarch Abraham.jpg
2.16 WCPD File:Sir Isaac Newton (1643-1727).jpg
2.17 WCPD File:Newton-Principia-Mathematica 1-500x700.jpg
2.18 WCPD File:Marshall's flax-mill, Holbeck, Leeds - interior - c.1800.jpg
2.19 WCPD File:Manuscrit déclaration des droits de l'homme et du citoyen.jpg
2.20 WCPD File:Huns in Italy by Checa.jpg
2.21 WCPD File:1843 Malte Brun Map of the Biblical Lands of the Hebrews (Egypt, Arabia, Israel, Turkey) - Geographicus - Hebreux-maltebrun-1837.jpg
2.22 WCPD File:Beginnings hist greece.jpg
2.23. WCPD File:Ancient Temple at Corinth engraving by William Miller after H W Williams.jpg
2.24. WCPD File:Ascetic asanas2.jpg; WCPDFile:Tin Hau buddah from afar.jpg; WCPD File:'Tirthankara', India, Mysore, Karnataka, 10th-11th century, bronze with silver content, Honolulu Academy of Arts.JPG

2.25. WCPD File:Solon writing laws for Athens.jpg
2.26. WCPD File:Homer British Museum.jpg; WCPD File:1904 Lawrence Alma-Tadema - The Finding of Moses.jpg; WCPD File:Krishna and Arjuna.jpg; WCPD File:Wrath of Achilles2.jpg
2.27. WCPD File:041A.Moses Breaks the Tables of the Law.jpg
2.28 WCPD File:Jacques-Louis David - Homer Reciting his Verses to the Greeks - WGA06120.jpg
2.29. WCPD File:Aryballos Macmillan.JPG
2.30. WCPD File:Relleu d'una màscara tràgica.JPG
2.31. WCPD File:The Phillip Medhurst Picture Torah 49. Going into the Ark. Genesis cap 7 v 7. Borcht.jpg
2.32. WCPD File:Banyans Yogis.jpeg
2.33. WCPD File:Confucius 02.png
2.34. File:Indo-European Migrations. Source David Anthony (2007), The Horse, The Wheel and Language.jpg: This file is licensed under the Creative Commons Attribution-Share Alike 4.0 International license. Indo-European Migrations. Source David Anthony (2007), The Horse, The Wheel and Language Date 31 January 2015, 12:54:30 Source Own work Author Joshua Jonathan
2.35. WCPD File:Dore joshua crossing.jpg
2.36. WCPD File:Rockcut ajanta cave 19.jpg
2.37. WCPD File:Media, Babylon and Persia - including a study of the Zend-Avesta or religion of Zoroaster, from the fall of Nineveh to the Persian war (1889) (14594576178).jpg
2.38. WCPD File:God Lord Shiva.jpg
2.39 File:Brooklyn Museum - Jain Manuscript Page.jpg , 1673: This image was uploaded as a donation by the Brooklyn Museum, and is considered to have no known copyright restrictions by the institutions of the Brooklyn Museum.
2.40. WCPD File:Rock painting-Confucius Meeting Lao Tzu (Rubbing).jpg.
2.41. File:Foxe's Christian martyrs of the world; the story of the advance of Christianity from Bible times to latest periods of persecution (1907) (14780749791).jpg. No known copright restrictions
2.42. WCPD File:NewYorkCityManhattanRockefellerCenter.jpg
2.43. WCPD File:Sumer.JPG
2.44. WCPD File:Die Gartenlaube (1891) b 025.JPG
2.45. WCPD File:'North Against South' by Léon Benett 67.jpg
2.46. WCPD File:Schinkel, Karl Friedrich - Gotische Kirche auf einem Felsen am Meer - 1815.jpg
2.47. WCPD File:Auguste Borget's oil on canvas painting 'An Indian Mosque on the Hooghly River near Calcutta', 1846.jpg
2.48. WCPD File:Wilberforce john rising.jpg
2.49. WCPD File:Paul Bril - Landscape with Roman Ruins - WGA03189.jpg
2.50. WCPD File:PeopleStormingTuileries.jpg
2.51. WCPD File:Foxe's Book of Martyrs - Tyndale.jpg
2.52. WCPD File:King-James-Version-Bible-first-edition-title-page-1611.png
2.53. WCPD File:Spinoza.jpg

Illustrations

2.54. WCPD File:Page077 Die letzte Bauernschlacht bei Frankenhausen am 25. Mai 1525.jpg
2.55. WCPD File:Mayflower compact.jpg
2.56. WCPD File:Karl Aspelin-Luther uppbränner den påfliga bullan.jpg
2.57. WCPD File:Hans Holbein, the Younger - Sir Thomas More - Google Art Project.jpg
2.58. WCPD File:Utopia.jpg
2.59. WCPD File:Thomas Muentzer.jpg
2.60. WCPD File:Levellers' Manifest.jpg
2.61. WCPD File:Gottfried Wilhelm von Leibniz.jpg
2.62. WCPD File:Newton's Method of Fluxions.jpg
2.63 WCPD File:Immanuel Kant.jpg
2.64 WCPD File:AdamSmith1790.jpg
2.65. Schopenhauer
2.66 WCPD Category:Claudio Monteverdi
2.67 WCPD File:Hegel.jpg
2.68. WCPD File:Hw-marx.jpg From: H.F. Helmolt (ed.): History of the World. New York, 1901.
2.69 WCPD File:Hobbes Locke Rousseau.jpg
2.70 WCPD File:Descartes3.jpg
2.71 WCPD File:ANICET-CHARLES-GABRIEL LEMONNIER A READING OF VOLTAIRE.jpg
2.72 WCPD File:Spherical pendulum Lagrangian mechanics.svg
2.73 WCPD File:Croce-Mozart-Detail.jpg
2.74 WCPD File:John Keats after J. Severn.jpg
2.75 WCPD File:Beethoven AlmanachDerMusikgesellschaft 1834.jpg
2.76.WCPD File:Thomas Paine.jpg
2.77. WCPD File:Dreadful Scene at Peterloo.png
2.78. WCPD File:Nietzsche.jpg
2.79. WCPD File:Pallas Athene c1539.jpg

Conclusion
WCPD Header File:Alessandro Magnasco - Banditti at Rest - WGA13845.jpg1. WCPD File:Aldrin Apollo 11.jpg

Appendix header File:Spera, Clemente and Magnasco, Alessandro - Mythological Figures among Ruins - 1690s.jpg

7. INDEX

A

abolitionism
  eonic emergence of 214
Abraham/Isaac meme 29
acceleration
  a devastating question
  and eonic transition 144
  cultural 111
adaptation 84
Advaita 256
After The Ice (Mithen)
Age of Revelation
  and the Axial Age 53
Akhenaton
  and Moses 16
alien analog 267
angelic memes 28
anti-modernism
  of Nietzsche 56
archaeology
  discovery of the Axial Age
Archaic Greece 73, 133, 142, 146
  Axial correlation
  evolution in jumps 140
  in The Origins of Greek Civilization 140
  stages of 142
Armstrong, Karen 55
atheism 56
atheism/theism
  stalemate with 267
Axial Age, The 7, 16, 49, 51, 54, 62, 73, 89, 260, 263
  and discontinuity 51
  and modernity 30
  and non-random patterns
  a second? 56
  as evidence of high-speed evolution 51
  clues shown by 62
  dates of 20, 64
  empirical basis 51
  macro aspect 20
  moving beyond sacrifice 29
  non-random pattern of
  scholarship of 38
Axial Greece
  and secularism 266
axis of history
  and the Axial Age 53
Axis points 64
  Axial Age as 'axis'

Index

in sequence 54

B

Ba Bird
  Egyptian soul concepts 28
Bazaz, Prem Nath
  on Buddhist revolution 161
being, function, will
  spiritual psychology 15
Bennett, J. G. 15, 21, 41, 262, 269
  and demiurgic powers 21
  his triad of 'being, function, will' 15
  on demiurgic powers 12
Bhagavad Gita, The
  and Brahmanism 161
Biblical Criticism 66, 148
Big History
  and Universal History 93
bodies of light 267
Buddhism 54, 82, 271
  and classical divide 145
  and Hinduism 161
  and Samkhya 164

C

Cambrian, The 62
causality
  system action and free action 99
Christian faith
  and universal history 9
Christianity 64, 73, 268, 269
  and Zoroastrianism 106
civilization
  evolution of 59
Classical Samkhya, An Interpretation of its History and Meaning (Larsen) 168
Climbing Mount Improbable (Dawkins) 75
Confucius
  in Axial period 129
consciousness
  ambiguity 260
cultural evolution 141, 143
  hyper-cultural evolution

Cyrus the Great 106

D

Darwin, Charles
Darwin debate 78, 82
Darwinism 20, 27, 62, 74, 85
  and natural selection 77
  assumptions of random evolution 67
  legacy of 74
Dawkins, Richard 75
  and atheism 35
demiurge, the 13, 28
  in Kant 12, 266
demiurgic powers 12, 42, 71, 261
  and avatars 19
  Bennett's definition 14
design argument, the 71, 264
  and evolution 66
  and evolution debate 13
determinism
  vs creativity 99
developmental biology
  and the eonic effect 78
dialectic, the
  and Samkhya 167
  in Kant 12
directionality 103
  and non-random evolution
  and teleology 11
  cyclical aspect 95
discrete/continuous models
  and macro effect 266
divinities
  and falsification 266
Dobzhansky, Theodosius
  meaning of evolution 7
Dynastic Egypt 54

E

early modern, the
  chronology 58
East and West 163
edge of space, the
  antinomy from Plato 12

as antinomy 262
Egypt
Elohim, the 13, 266
end of history 111
   Fukuyama on 111
   in Hegel's philosophy of history
End of History and the Last Man, The
     (Fukuyama) 111
Enlightenment, The 58, 112
eonic effect, The 192, 221
   bottom line on evolution
   dose of empiricism
     Old Testament 145
   philosophy of history 100
eonic emergence
   Buddhism and Samkhya 167
   TP3 zoom targets 203
eonic evolution
   of religion 109
eonic sequence 143
   history of Israel 153
eonic transition 158
  basic thesis
  history of Israel 154
  sources of Chinese tradition 168
evo-devo 78
evolution 34
   and history 58
   and natural selection 77
   climbing Mt. Improbable 75
   defining 65
   end of history 111
   Kant's Challenge 100
   Lamarck 18
   meaning of 8
   metaphysics of 82
   non-random 58
   of philosophy of history 100
   Samkhya 167
Evolution From Space (Hoyle) 77
Exile, The 143
   onset of backward looking 145
   origins of eschatology 107
   Persians and Zoroastrianism 106

F

feedback
   in evolutionary sequences 67
Feuerbach, Ludwig 264
First World War, The 56
Fisher's Lament 62
'food for god' rackets 27
four-dimensional black box 269
free agency
   and system action 269
freedom
   and Kant 269
freedom's causality 225
Free Will
   and Kant 224
French Revolution 216, 218
   modern transition 203
frequency hypothesis
   and a frequency deduction 97
Frontier Effect, The 59
Fukuyama, Francis 111

G

Gaia
   and civilization 266
   Axial Age and 46
Gilgamesh
Gita As It Was, The (Sinha)
   pre-theistic Gita 164
Goldilocks Enigma, The 93
Great Explosion, The 86
   and human evolution 86
Great Transition, The 57, 60
Greece
   Archaic period
   Bertrand Russell on 123
   spectrum of the spiritual 156
Greek Archaic, The 194
   sudden emergence of 133

H

Higher Power, A
   in history 263

Index

Hinayana
  and Mahayana 226
historical inevitability 98
historical model 57
historicism
  of freedom 264
  two types of 96
history
  and evolution 84
  a science of 84
  philosophy of 62
History of the World (Roberts) 192
History of Western Philosophy, A (Russell) 123
Hodgson, Marshall 193
Hoffman, Michael 188
Holocaust, The 56
Homer 272
homo sapiens
  and consciousness 261
Hoyle, Fred 67, 77
  probability and evolution 67
human consciousness
  and 'soul' 19
human emergence
  theories of 17
hurricane argument 79
Huxley, T. H. 74
hypermechanical 266, 268
hypernomic, the 27
  and Samkhya 12, 262
hyponomic
  and autonomic, hypernomic 12, 260

I

Idea For A Universal History (Kant) 100, 111
idealism
  transcendental 61
idea of progress
  Fisher's Lament 96
IHVH 13, 25, 32, 43, 230, 66, 68, 231, 261
  nameless 'god' 263
  silence as to divine name

imperialism
  Roman or American empire 118
India
  Upanishadic era
Intelligent Design 74
  movement of 74
Iron Cage, The 72
Islam 73, 269
Israel
  eonic periodization 153
  in the Axial Age 129
  t-stream and e-sequence 153
Israel/Judah 264, 267
Israel/Persia
  and Axial Age 273

J

Jainism 54, 73
Jaspers, Karl , 9, 49, 70
  and Axial Age 129
  and Kant 16
  Eurocentrism 22
Jehovah 68, 268
  and IHVH 264
Judaism 269

K

Kant, Immanuel 13, 55, 80, 91
Kant's Challenge 7, 24, 62, 99, 271

L

Lamarck, Jean-Baptiste 18, 87
Lenoir, Timothy 83
  on the teleomechanists 83
liberalism 215

M

MacNeill, William 192
macro-action
  and micro-action 97
macro effect, the 7, 89
macroevolution 78, 81, 86
macro pattern, the 266

macro processes
  and evolutionary drivers 67
macrosequence. 46
Mahavir 265
Mahayana 71, 269
materialism 120
Mayans
  Axial Age synchrony 31
meditation
  and human psychology 261
metanarratives 96
microevolution 81, 86
Middle Ages 154
modernism 110
modernity 115, 196
  and religion 32
  a second Axial Age? 187
  rise of 52
  sudden explosion of 112
  vs the postmodern 112
molecular biology 76
monotheisim 71
Monotheism 261, 264
monotheistic myth, the
  of Israel 20
Moses
  as historical figure 109
Mount Improbable
  climbing 75

N

natural selection 74, 77, 86
  incoherence of 8
  theory of 62
  vs design argument 34
natural teleology
  and Kant 25, 83
Nature's Secret Plan 26
Nehru, Jawaharlal 163
Neolithic, The
Newton, Isaac 78
Nietzsche, Friedrich 22, 70
  and Spengler 116
non-random evolution

detecting 58
non-random, the
  patterns in world history
noumenal, the
  and phenomenal 95

O

observation
  limits of 79
Old Testament, The 49, 145, 261, 266, 267
  and Axial Age , 51
  as eonic data 145
  in the Axial period
Omega Point 111
Origin and Goal of History, The 9
Origin and Goal of History, The (Jaspers) 120
  implications in title

P

paganism
  and sacrifice 16
periodization
  rise of modern 194
Persia
  and Zoroastrianism 129
Persian Empire
  and Book of Daniel 106
  and Zoroastrianism 106
  Judaic Exile 107
  transition from empire to religion 155
philosophy of history 98
photo finish test 84
Polanyi, Karl
  personal knowledge 72
polytheism 268
pop theism 264
positivism 56
postcapitalism 235
postmodernism 96
  and Fisher s Lament 96
post-religion
  trend toward 30
Predynastic Egypt (Hoffman) 188

Index

Prehistoric Europe (Stern) 188
Prophets, The
  and classical phase
    in Old Testament 264
  periodization of transition 155
punctuated equilibrium 64
  and speciation

R

random evolution 67
  problems with 67
randomness
  eonic effect evidence of non-random
    Fisher's Lament 94
  world history flunks test of
randomness test
  world history flunks
reductionism 60
reformation
  and Old Testament 45
Reformation, The 111, 264, 270
  early modern period 191
relative beginnings 141
  in macrosequence 23
relative transformation
  and monotheism 109
religion
  and Buddhism 163
  Samkhya tradition 164
Rethinking World History (Hodgson) 193
revolution 143, 155, 163, 188, 189
  and Exodus myth 155
rise of civilization
  timeline 33
Rise of the West, The (MacNeill) 192
Role of the Bhagavad Gita in Indian History, The (Bazaz) 161
Rome 59
Rosetta Stone, The
  and Champollion 92
Rousseau, Jean Jacques
  and Kant 211
Russell, Bertrand
  on Greeks 123

S

sacrifice
  and Aryan Vedism 29
  history of religion 29
Samkhya 12, 42, 164, 260
  early history of 164
  level 12, 6, 3 268
Schopenhauer, Arthur 15, 262
  the Will in Nature 269
science of history 84
Scientific Revolution, The 111
  as eonic emergent 203
scientism 25
secularism 20, 52, 112, 187
  meaning of 112
self-consciousness
  vs. consciousness 261
shaman, the 19
Shiva Seal, The 27
Sinha, Phulgenda
  quest for original Gita 164
slavery 214
Solon 155
Spengler, Oswald
  and declinism 115
Spinoza, Benedict 267
'spiritual' powers
  in nature 12
SPR-MAT-X 14, 268
  and demiurgic powers 267
  and hypernomic 260
stream and sequence 59
  and macro model 23, 37
sufficient reason
  principle of 64
Sufism 267
Sumer 54, 59, 140, 193
synchronism 145
synchronous emergence
  and The Axial Age , 51
synchrony
  parallel emergence , 51
system action
  and free action

systems analysis
  black boxes and 269

T

technostream
  Chinese science 170
Teertankers 14, 73, 265
teleology 78
  and non-random evolution 67
  detecting 67
Ternate letter
  and Darwin 63
The Strategy of Life (Lenoir) 83
Third Antinomy, The
  of Kant 97
Toynbee, Arnold 116
tragedy
  and teleology 152
'triple play', the 7
turning points
  rise of the modern 191

U

Universal History 107, 96, 108
Upanishads, The
  Axial Age timing of 130
Ur 59

V

Van Doren Stern, Philip 188
Visions of a Ghostseer
  Kant on metaphysics 61
void, the
  and buddhsim 263

W

Wallace, Alfred 63
Western Transmutation
  See Hodgson 193
world history
  and sudden emergence
  evidence of evolution , 51
  flunks randomness test

Z

Zarathustra 120, 122
  millenialism 106
Zoroastrianism 155

 The author is a poet, student of Classics, Mathematics, World History, and Eastern Religion. World traveller, volunteer in Peace Corps, student of Sufism, Jainism, and the religious archaeology of India. Author of *World History and The Eonic Effect*, with multiple websites and blogs, including *Darwiniana*, the evolution blog, by nemo/nemini (history-and-evolution.com/nemonemini), and six other blogs and websites linked to each other.

The book was set in Minion Pro, Garamond, and Broadway fonts, using Adobe Indesign and Photoshop CS5.5.

CPSIA information can be obtained
at www.ICGtesting.com
Printed in the USA
LVHW050849170422
716425LV00008B/558